FATIGUE DESIGN

Life Expectancy of Machine Parts

Eliahu Zahavi

with Vladimir Torbilo

CRC Press
Taylor & Francis Group
Boca Raton London New York

CRC Press is an imprint of the
Taylor & Francis Group, an **informa** business

CRC Press
Taylor & Francis Group
6000 Broken Sound Parkway NW, Suite 300
Boca Raton, FL 33487-2742

First issued in paperback 2019

ISBN-13: 978-0-8493-8970-2 (hbk)
ISBN-13: 978-0-367-40133-7 (pbk)
Library of Congress Card Number 96-14021

Library of Congress Cataloging-in-Publication Data

Zahavi, Eliahu.
    Life expectancy of machine parts : fatigue design / Eliahu Zahavi, Jesa H. Kreiner.
        p.   cm.
    "A Solomon Press book."
    Includes bibliographical references and index.
    ISBN 0-8493-8970-4 (alk. paper)
    1. Machine parts. 2. Materials--Fatigue. I. Kreiner, Jesa H. II. Title.
    TJ243.Z29   1996
    621.8'2--dc20

                                                                    96-14021
                                                                    CIP

CONTENTS

Chapter 6   Surface Integrity and Fatigue   **183**

Chapter 7   Fatigue Life Improvement   **239**

PREFACE

To make the book as applicable as possible, the initiator of this work took an interdisciplinary approach, combining in one volume the two related fields concerned with fatigue design: theoretical foundation of fatigue analysis and study of behavior of materials in manufacture of machine parts. The solved fatigue failure cases, taken from real situations, illustrate the inter-dependency of the two. The theoretical foundation and the cases (chapters 1, 2, 3, 4, 5, and 8) written by Eliahu Zahavi, cover a wide range of fatigue problems: from simple to nonlinear in multi-axial space. On the other hand, the thorough coverage of surface phenomena in chapters 6 and 7 puts in proper perspective the other perils that must be considered in designing the machine parts for fatigue life, and is the contribution of Vladimir Torbilo.

Practical constraints prevent us from acknowledging here personally all those who helped us accomplish what we set out to do, not least our families for their devotion and forbearance. Among the few we did single out, a sincere appreciation goes to Professor A. Berkowits, of Technion, Israel, for the review of the chapters on fatigue theories and for the kind permission to include his latest maxims here. To each one of the reviewers of case studies, Emanuel Liban, Simha Bar-Nefi and David Barlam, our gratitude for their comments and professional advice.

Joe Claypool and Bob Stern (CRC Press) deserve our special mention for their patience, understanding, and publishing know-how. We are grateful to Mary Russell, our copyeditor, for helping to make this work so much more readable; and to Nitza Yogev, a talented illustrator, for most of the drawings in the book.

This is our second book produced under the management of The Solomon Press, and our deepest indebtedness goes to Sidney Solomon and Raymond Solomon for their professional skills, continued encouragement and personal involvement in every aspect of this undertaking.

1

INTRODUCTION

If it were feasible to endow an individual who sets out to design machinery with all the knowledge that engineering design directly or indirectly relies on, there is no doubt that such an individual would need no assistance other than from his creative mind. The body of knowledge that machine design relies on includes a great number of disciplines. A few of these are: solid mechanics, including the elasticity and plasticity theories; metallurgy with its many faceted applications; manufacturing processes and surface effects; and numerical applications, especially the finite element method. To compromise with reality, the design engineer relies on the experience and fruits of research of other specialists.

A machine design process is comprised of many different aspects. The objective of this book is to delve into that phase of the process which is concerned with fatigue. Namely, creating machine parts, operating under fluctuating loading, with no risk of an unexpected failure. The challenge in creating a part with a predictable life expectancy is great because it is a known fact that a large number of machine parts fail due to fatigue. Some specialists claim fatigue failure reaches up to 80 percent, others claim the percentage at one hundred.

The history of fatigue design goes back to the middle of the nineteenth century, marked by the beginning of industrial revolution and, in particular, the advent of railroads in central Europe. The first known investigators concerned with fatigue phenomena were designers of axles for locomotives and wagons. Wohler's experiments (1858)[1] with axles were the first known laboratory tests with the objective to derive and quantitatively describe the limits of fatigue. This was followed by more elaborate analyses of stresses and their effect on fatigue by Gerber,[2] Goodman[3] and others. Continuous efforts of researchers in the twentieth century have given a new impetus to the development of theories, such as the effects of plastic deformation on fatigue—resulting in the strain method discovered by Manson[4] and Coffin.[5] In parallel, the theory of crack propagation started by Griffith (1921)[6] was continued by Paris[7] and others. Research accomplishments of Morrow,[8] Socie[9] and their followers brought the state of fatigue analysis to the present day level. Of special importance are works published by SAE Fatigue Design and Evaluation Committee.[10] The challenges of ever growing modern technology demand a continuation of research efforts toward the more exact prediction of fatigue lives.

SCOPE OF THE BOOK

The topics in the book are grouped into four distinctive parts as follows: fundamentals of solid mechanics, Chapter 2; methods of fatigue analysis, Chapters 3, 4 and 5; impact of production processes upon the fatigue, Chapters 6 and 7; and design cases, Chapter 8.

An analysis of fatigue phenomena requires a thorough understanding of the interaction of stresses and strains within the body under loading. We address this topic in Chapter 2, presenting the necessary fundamentals of the theories of elasticity and plasticity. The advent of digital computers has made possible the application of these theories by means of numerical methods. A description follows of the most practical method as applied to the stress and strain analysis—the finite element method. Solutions to nonlinear problems encountered in fatigue design are reached by a special adaptation of this method. Equipped with this knowledge, the reader can proceed to the topic of fatigue analysis.

The organization of the second section follows the chronological development of the methods of fatigue analysis, comprising three known approaches—the stress method, the strain method and crack propagation theory, respectively. Chapter 3 is devoted to the description of the basic approach, the stress method. The method stems from the assumption that a machine part under fluctuating loading undergoes elastic deformation only. The method was developed for designing of parts with an infinite life introducing a safety factor as a measure. With time it was extended to include the design of parts with a limited life, a given life expectancy. The strain method, Chapter 4, is based on a more profound approach, taking into consideration the occurrence of plastic deformation, as well. As a result of an extended theoretical base and better accuracy, the method has found wide application in the automotive and aviation industries. The theory that deals with the final stage of fatigue, the crack propagation, is described in Chapter 5. The theory considers the fatigue life of the machine part—the period from an initiation of a crack to its causing a fracture and final failure of the machine part. The period following the initial crack which continues to grow to a final fracture of a machine part, is called the destruction period.

An overview is needed to understand the relation of the first two methods (the stress method and the strain method) to the third one, the crack propagation method. The first two differ fundamentally in defining the limits of fatigue life. The stress method includes the destruction period, while the strain method limits the life up to the presence of a visible crack up to 3–6 mm in length. Accordingly, in computation, the former method uses parameters taken from tests which include the destruction period, while the latter uses test data that limit the fatigue life to the onset of the crack.

Having created an optimal design of a machine part with the help of the appropriate fatigue design theories, presented above, the manufactured part may still meet with hidden perils of failure that must be foreseen during the design. To minimize such risks that stem from manufacture and later from the adverse conditions in operation, the discussion here relates to certain aspects of the production processes. One of the critical influences on the fatigue life is the condition of the machined surface. A fatigue failure begins in most cases at the surface because the surface layer bears the greatest load and is exposed to environmental effects. Part three of the book is concerned with the surface layer quality, Chapter 6, and fatigue life improvement, Chapter 7.

The surface is conditioned by the production process. The finish operation creates surface roughness and, due to plastic deformation and heating, structural changes and residual stresses arise. An unsuitable or faulty surface roughness can cause crack initiation and propagation, while unfavorable residual stresses impair the fatigue life. It is imperative therefore to consider the available preventive measures at the design stage. A number of methods to improve the fatigue strength of a machine part are presented.

The last section of the book (Chapter 8) illustrates three design cases all experiencing fatigue problems. The diversified cases were chosen from real experiences concerning aircraft, military equipment and oil refineries. As is shown by the analysis, in all three cases no fault could be attributed to a negligent design of the corresponding parts. This fact highlights the presence of perilous factors that can not be predicted at the design stage.

The illustrated solutions in this part include the application of the theories of fatigue analysis discussed elsewhere in the book and demonstrate the use of the finite element method. The data fed into the solution are mostly based on the available professional literature and assumptions, without direct involvement in the cases. The solutions used two computer programs, ANSYS and MSC/NASTRAN, as per printouts. These solutions are to be considered as illustrative examples only, based on our limited understanding of the conditions leading to failure.

NOMENCLATURE

The careful reader will note different nomenclature throughout the book, as used in conjunction with different analytical methods. This applies especially to the designation of stresses: letter S is used in the stress method while elsewhere the stress is designated by σ. The reason for this discrepancy is due to a tradition of the stress method throughout its long history. This nomenclature continues to be the one in use by design engineers in application of this method. (It is still covered by SAE[11] and ASTM[12] standards.)

REFERENCES

1. Wohler, A. 1858. "Uber die Festigkeitsversuche mit Eisen und Stahl," *Z. Bauwesen.* 8, 641.
2. Gerber, W. 1874 "Bestimmung der zulassigen Spannungen in Eisenkonstruktionen," *Z. Bayer, Arch, Ing. Ver,* 101.
3. Goodman, J. 1899. *Mechanics Applied to Engineering.* London: Longmans.
4. Manson, S.S. 1954. "Behavior of Materials under Conditions of Thermal Stress," *NACA Tech. Note 2933.*
5. Coffin, L.F. 1954. "A Study of Cyclic-Thermal Stresses in a Ductile Material," *Trans. ASME.* 76, 931–950.
6. Griffith, A.A. 1921. "The Phenomena of Rupture and Fracture in Solids," *Phil. Trans. Roy. Soc.* A 221, 163–197.
7. Paris, P.G. 1963. "The Fracture Mechanics Approach to Fatigue," Proc. 10th Sagamore Conference. Syracuse University Press.
8. Morrow, J. 1965. "Cyclic Plastic Strain Energy and Fatigue of Metals," *Internal Friction, Damping and Cyclic Plasticity,* ASTM STP 378, 45–87.
9. Socie, D.F. 1977. "Fatigue Life Prediction Using Local Stress-Strain Concepts," *Experimental Mechanics.* 17, No, 2.
10. "Fatigue Under Complex Loading." 1977. Wetzel, R.M. ed., Warrendale, PA: Society of Automotive Engineers. P.G.
11. "SAE Handbook." 1989. Warrendale, PA: Society of Automotive Engineers.
12. "Annual Book of ASTM Standards. 1980. " Part 10, Philadelphia, PA: ASTM.

2

SOLID
MECHANICS

The fatigue failure of a machine part under loading is contingent first and fore-most on the stresses and strains within. The theoretical foundation used for fatigue analysis involves two attributes created by stresses: an elastic state existing below the yield point of the material and a plastic state existing above it. Our introduction here of corresponding theories of elasticity and plasticity is of a limited scope and is directed mainly to enable one to understand the computational methods used in fatigue analysis.

2.1
ELASTICITY

2.1.1 Stress

When external loading is applied to a body, the body becomes deformed and inter-nal stresses occur. To analyze the stresses occurring within elastic conditions, we make the following assumptions: (a) the material is homogeneous and isotropic; (b) the deformations are small so that the body dimensions are substantially unchanged; (c) the body regains its original form when the applied load is removed; and (d) the strains are infinitesimal, permitting an assumption of linear elasticity.

Consider a body subject to an applied load in a Cartesian coordinate system within elastic domain. We begin with investigation of an elemental cube, Figure 2.1. Each face

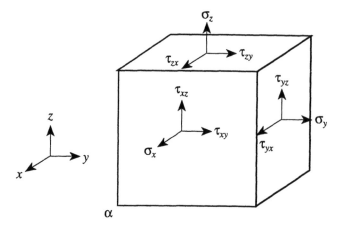

Figure 2.1 Elemental cube subjected to stresses in Cartesian coordinate system.

of the cube is subjected to stresses: normal stress components perpendicular to a surface and shear components tangent to a surface. See figure. Consider the following nine stress components—three normal, $\sigma_x, \sigma_y, \sigma_z$, and six shear, $\tau_{xy}, \tau_{yz}, \tau_{zx}, \tau_{xz}, \tau_{zy}$ and τ_{yx}—which form the stress tensor

$$\sigma_{ij} = \begin{pmatrix} \sigma_x & \tau_{xy} & \tau_{xz} \\ \tau_{yx} & \sigma_y & \tau_{yz} \\ \tau_{zx} & \tau_{zy} & \sigma_z \end{pmatrix} \tag{2.1}$$

The elemental cube is in static equilibrium and therefore the sum of moments relative to edge AB must be zero, that is,

$$\begin{aligned} \tau_{yx} &= \tau_{xy} \\ \tau_{xz} &= \tau_{zx} \\ \tau_{zy} &= \tau_{yz} \end{aligned} \tag{2.2}$$

Equations (2.2) are based upon the assumption that there are no coupling effects of thermal or other character and therefore the state of stress can be expressed in terms of six components.

Principal stresses

Consider further the elemental cube, Figure 2.2. Assume that a plane intersects the cube at an inclination defined by angles (n,x), (n,y), and (n,z) where n represents a normal direction to the plane. Three stress components act upon the section:

$$\begin{aligned} s_x &= \sigma_x \cos(n,x) + \tau_{yx} \cos(n,y) + \tau_{zx} \cos(n,z) \\ s_y &= \tau_{xy} \cos(n,x) + \sigma_y \cos(n,y) + \tau_{zy} \cos(n,z) \\ s_z &= \tau_{xz} \cos(n,x) + \tau_{yz} \cos(n,y) + \sigma_z \cos(n,z) \end{aligned} \tag{2.3}$$

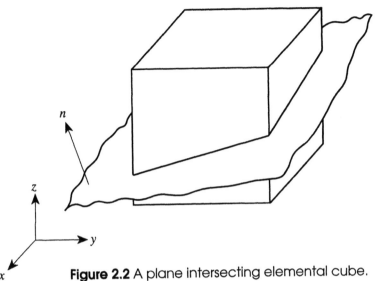

Figure 2.2 A plane intersecting elemental cube.
n represents a normal to the plane.

Components s_x, s_y and s_z form the stress vector s. The cube can be intersected at an inclination so that vector $s = s_n$ becomes normal to the intersecting plane. Normal n to such plane is called principal direction. Stresses acting in a principal direction are called principal stresses. As shown below, there are three principal axes for each state of stress. (On planes perpendicular to principal axes there are no shear stresses.)

Let us denote a principal stress by σ. When direction n is a principal direction, we have

$$s_x = \sigma \cos(n,x), \quad s_y = \sigma \cos(n,y), \quad s_z = \sigma \cos(n,z) \qquad [2.4]$$

Inserting Equations (2.5) into Equations (2.4) we obtain

$$\left(\sigma_x - \sigma\right)\cos(n,x) + \tau_{yx}\cos(n,y) + \tau_{zx}\cos(n,z) = 0 \qquad [2.5]$$
$$\tau_{xy}\cos(n,x) + \left(\sigma_y - \sigma\right)\cos(n,y) + \tau_{zy}\cos(n,z) = 0$$
$$\tau_{xz}\cos(n,x) + \tau_{yz}\cos(n,y) + \left(\sigma_z - \sigma\right)\cos(n,z) = 0$$

These equations can be solved for $\cos(n,x)$, $\cos(n,y)$, and $\cos(n,z)$ only if the following determinant equals zero,

$$\begin{vmatrix} \left(\sigma_x - \sigma\right) & \tau_{yx} & \tau_{zx} \\ \tau_{xy} & \left(\sigma_y - \sigma\right) & \tau_{zy} \\ \tau_{xz} & \tau_{yz} & \left(\sigma_z - \sigma\right) \end{vmatrix} = 0 \qquad [2.6]$$

Opening the determinant produces a cubic equation in terms of σ

$$\sigma^3 - \left(\sigma_x + \sigma_y + \sigma_z\right)\sigma^2 + \left(\sigma_x\sigma_y + \sigma_y\sigma_z + \sigma_z\sigma_x - \tau_{yz}^2 - \tau_{zx}^2 - \tau_{xy}^2\right)\sigma \qquad [2.7]$$
$$- \left(\sigma_x\sigma_y\sigma_z + 2\tau_{yz}\tau_{zx}\tau_{xy} - \sigma_x\tau_{yz}^2 - \sigma_y\tau_{zx}^2 - \sigma_z\tau_{xy}^2\right) = 0$$

Solving Equation (2.7) for σ, we obtain three roots, principal stresses σ_1, σ_2, and σ_3. In terms of principal stresses the stress tensor, Equation (2.1), becomes

$$\sigma_{ij} = \begin{pmatrix} \sigma_1 & 0 & 0 \\ 0 & \sigma_2 & 0 \\ 0 & 0 & \sigma_3 \end{pmatrix} \tag{2.8}$$

Inserting σ_1, σ_2, and σ_3 into the system of Equations (2.5), we obtain three sets of values of $\cos(n,x)$, $\cos(n,y)$, and $\cos(n,z)$ which define three perpendicular principal directions.

Stress invariants   Let us introduce expressions I_1, I_2, and I_3 defined as follows

$$I_1 = \sigma_x + \sigma_y + \sigma_z = \sigma_1 + \sigma_2 + \sigma_3 \tag{2.9}$$
$$I_2 = \sigma_x\sigma_y + \sigma_y\sigma_z + \sigma_z\sigma_x - \tau_{yz}^2 - \tau_{zx}^2 - \tau_{xy}^2 = \sigma_1\sigma_2 + \sigma_2\sigma_3 + \sigma_3\sigma_1$$
$$I_3 = \sigma_x\sigma_y\sigma_z + 2\tau_{yz}\tau_{zx}\tau_{xy} - \sigma_x\tau_{yz}^2 - \sigma_y\tau_{zx}^2 - \sigma_z\tau_{xy}^2 = \sigma_1\sigma_2\sigma_3$$

whereby

$$\sigma^3 - I_1\sigma^2 + I_2\sigma - I_3 = 0 \tag{2.10}$$

I_1, I_2, and I_3 are stress invariants of the stress tensor, Equation (2.1). They are independent of the coordinate system (x,y,z).

Mohr diagram   Assume the stresser in z-direction to be zero, that means $\sigma_z = \tau_{zx} = \tau_{yz} = 0$. Let us denote the directional cosines of the plane intersecting the cube as

$$\cos(n,z) = 0 \quad \text{and} \quad \cos(n,z) = \sin(n,y) = \cos\alpha \tag{2.11}$$

As a consequence, the problem becomes a two-dimensional one. Equation (2.7) takes the form

$$\sigma^2 - (\sigma_x + \sigma_y)\sigma + (\sigma_x\sigma_y - \tau_{xy}^2) = 0 \tag{2.12}$$

The two principal stresses, σ_1 and σ_2, equal

$$\sigma_1 = \frac{\sigma_x + \sigma_y}{2} + \sqrt{\left(\frac{\sigma_x - \sigma_y}{2}\right) + \tau_{xy}^2} \tag{2.13}$$

$$\sigma_2 = \frac{\sigma_x + \sigma_y}{2} - \sqrt{\left(\frac{\sigma_x - \sigma_y}{2}\right) + \tau_{xy}^2}$$

The Mohr circle shown in Figure 2.3(a) reflects Equations (2.13) and illustrates graphically the dependence between the stresses and the inclination angle α within the plane

(1,2). We may obtain equivalent Mohr circles for planes (2,3) and (3,1). A superposition of the three circles provides a Mohr diagram for stresses in three dimensions, Figure 2.3(b). (It is assumed that $\sigma_1 > \sigma_2 > \sigma_3$.)

Principal shear stresses   From Mohr circle, Figure 2.3(a), it follows that the shear stress in plane (1,2) equals

$$\tau = \pm \frac{|\sigma_1 - \sigma_2|}{2} \sin 2\alpha$$

[2.14]

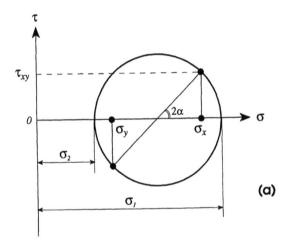

(a)

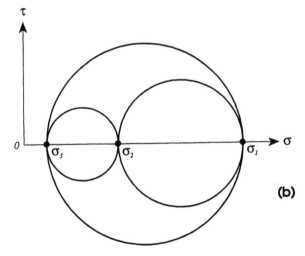

(b)

Figure 2.3 Mohr diagrams: (a) Mohr circle showing the dependence between stresses and inclination angle α within the plane (1,2); (b) Mohr circles for a three-dimensional body.

The maximum value of shear stress is equal to the radius $(\sigma_1 - \sigma_2)/2$. Correspondingly, for three-dimensional stresses, the three maximum shear stresses are

$$\tau_{12} = \frac{|\sigma_1 - \sigma_2|}{2}, \quad \tau_{23} = \frac{|\sigma_2 - \sigma_3|}{2}, \quad \tau_{31} = \frac{|\sigma_3 - \sigma_1|}{2} \qquad [2.15]$$

and τ_{12}, τ_{23} and τ_{31} are called principal shear stresses.

Stress deviator tensor

The stress tensor, Equation (2.1), can be expressed as a sum

$$\sigma_{ij} = \begin{pmatrix} \sigma_m & 0 & 0 \\ 0 & \sigma_m & 0 \\ 0 & 0 & \sigma_m \end{pmatrix} + \begin{pmatrix} \sigma_x - \sigma_m & \tau_{yx} & \tau_{zx} \\ \tau_{xy} & \sigma_y - \sigma_m & \tau_{zy} \\ \tau_{xz} & \tau_{yz} & \sigma_z - \sigma_m \end{pmatrix} \qquad [2.16]$$

where σ_m is a mean stress defined by the equation

$$\sigma_m = \frac{1}{3}\left(\sigma_x + \sigma_y + \sigma_z\right) = \frac{1}{3}\left(\sigma_1 + \sigma_2 + \sigma_3\right) \qquad [2.17]$$

The first tensor on the right side of Equation (2.16) is called spherical tensor and represents hydrostatic pressure. The second tensor is a stress deviator tensor, representing deviations from the mean stress, expressed as

$$\sigma_{ij} = \begin{pmatrix} \sigma_x - \sigma_m & \tau_{yx} & \tau_{zx} \\ \tau_{xy} & \sigma_y - \sigma_m & \tau_{zy} \\ \tau_{xz} & \tau_{yz} & \sigma_z - \sigma_m \end{pmatrix} = \begin{pmatrix} \sigma_1 - \sigma_m & 0 & 0 \\ 0 & \sigma_2 - \sigma_m & 0 \\ 0 & 0 & \sigma_3 - \sigma_m \end{pmatrix} \qquad [2.18]$$

$$= \begin{pmatrix} \dfrac{2\sigma_1 - \sigma_2 - \sigma_3}{3} & 0 & 0 \\ 0 & \dfrac{2\sigma_2 - \sigma_3 - \sigma_1}{3} & 0 \\ 0 & 0 & \dfrac{2\sigma_3 - \sigma_1 - \sigma_2}{3} \end{pmatrix} = \begin{pmatrix} s_1 & 0 & 0 \\ 0 & s_2 & 0 \\ 0 & 0 & s_3 \end{pmatrix}$$

The invariants of the stress deviator tensor are

$$\begin{aligned} J_1 &= s_1 + s_2 + s_3 = 0 \\ J_2 &= s_1 s_2 + s_2 s_3 + s_3 s_1 \\ J_3 &= s_1 s_2 s_3 \end{aligned} \qquad [2.19]$$

The middle invariant, J_2, may be expressed in an expanded form

$$J_2 = s_1 s_2 + s_2 s_3 + s_3 s_1 = -\frac{1}{6}\left[\left(\sigma_1 - \sigma_2\right)^2 + \left(\sigma_2 - \sigma_3\right)^2 + \left(\sigma_3 - \sigma_1\right)^2\right] \qquad [2.20]$$

which is referred to again in Section 2.1.5.

2.1.2 Deformation and strain

A body subject to stresses undergoes deformation in which the internal points are displaced. The deformation at each point is measured by a displacement per unit length which is called strain. In order to derive the correlation between stresses and deformation, we first must define the strain in more exact terms. A mathematical derivation of the strain tensor follows.

Each point within the body undergoing a deformation has a displacement vector

$$\{\mathbf{u}\} = \left\{\begin{array}{c} u \\ v \\ w \end{array}\right\} \qquad [2.21]$$

which is a function of location. The differential of the displacement vector comprises the following three components

$$du = \frac{\partial u}{\partial x}dx + \frac{\partial u}{\partial y}dy + \frac{\partial u}{\partial z}dz \qquad [2.22]$$

$$dv = \frac{\partial v}{\partial x}dx + \frac{\partial v}{\partial y}dy + \frac{\partial v}{\partial z}dz$$

$$dw = \frac{\partial w}{\partial x}dx + \frac{\partial w}{\partial y}dy + \frac{\partial w}{\partial z}dz$$

The above components can be presented in the vectorial form

$$\left\{\begin{array}{c} du \\ dv \\ dw \end{array}\right\} = \left(\begin{array}{ccc} \dfrac{\partial u}{\partial x} & \dfrac{\partial u}{\partial y} & \dfrac{\partial u}{\partial z} \\ \dfrac{\partial v}{\partial x} & \dfrac{\partial v}{\partial y} & \dfrac{\partial v}{\partial x} \\ \dfrac{\partial w}{\partial x} & \dfrac{\partial w}{\partial y} & \dfrac{\partial w}{\partial z} \end{array}\right) \left\{\begin{array}{c} dx \\ dy \\ dz \end{array}\right\} \qquad [2.23]$$

The first term on the right side of Equation (2.23) is a tensor

$$e_{ij} = \begin{pmatrix} \dfrac{\partial u}{\partial x} & \dfrac{\partial u}{\partial y} & \dfrac{\partial u}{\partial z} \\[2mm] \dfrac{\partial v}{\partial x} & \dfrac{\partial v}{\partial y} & \dfrac{\partial v}{\partial z} \\[2mm] \dfrac{\partial w}{\partial x} & \dfrac{\partial w}{\partial y} & \dfrac{\partial w}{\partial z} \end{pmatrix} \qquad [2.24]$$

Let us break down the above tensor into three component tensors as follows

$$e_{ij}^{(1)} = \begin{pmatrix} \dfrac{\partial u}{\partial x} & 0 & 0 \\[2mm] 0 & \dfrac{\partial v}{\partial y} & 0 \\[2mm] 0 & 0 & \dfrac{\partial w}{\partial z} \end{pmatrix} \qquad [2.25]$$

$$e_{ij}^{(2)} = \frac{1}{2}\begin{pmatrix} 0 & \dfrac{\partial u}{\partial y}+\dfrac{\partial v}{\partial x} & \dfrac{\partial u}{\partial z}+\dfrac{\partial w}{\partial x} \\[2mm] \dfrac{\partial v}{\partial x}+\dfrac{\partial u}{\partial y} & 0 & \dfrac{\partial v}{\partial z}+\dfrac{\partial w}{\partial y} \\[2mm] \dfrac{\partial w}{\partial x}+\dfrac{\partial u}{\partial z} & \dfrac{\partial w}{\partial y}+\dfrac{\partial v}{\partial z} & 0 \end{pmatrix} \qquad [2.26]$$

and

$$e_{ij}^{(3)} = \frac{1}{2}\begin{pmatrix} 0 & \dfrac{\partial u}{\partial y}-\dfrac{\partial v}{\partial x} & \dfrac{\partial u}{\partial z}-\dfrac{\partial w}{\partial x} \\[2mm] \dfrac{\partial v}{\partial x}-\dfrac{\partial u}{\partial y} & 0 & \dfrac{\partial v}{\partial z}-\dfrac{\partial w}{\partial y} \\[2mm] \dfrac{\partial w}{\partial x}-\dfrac{\partial u}{\partial z} & \dfrac{\partial w}{\partial y}-\dfrac{\partial v}{\partial z} & 0 \end{pmatrix} \qquad [2.27]$$

whereby

$$e_{ij} = e_{ij}^{(1)} + e_{ij}^{(2)} + e_{ij}^{(3)} \qquad [2.28]$$

Figure 2.4 illustrates the meaning of the three tensors. It shows a deformed rectangle which is a projection of the elemental cube on *xy*-plane. Consider the illustrated body

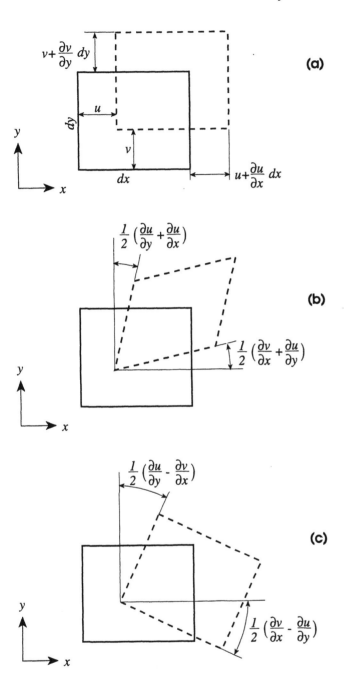

Figure 2.4 Deformation of a body under load:
(a) volumetric deformation; (b) deformation
of the form; (c) pure rotation.

deformation as composed of three independent acts—a volumetric deformation, a deformation of the form, and a pure rotation. Similar projections can be drawn on the yz- and zx-planes. Investigating the three tensors, Equations (2.25), (2.26) and (2.27), the correlation of each tensor to a corresponding act in the figure becomes apparent. Tensors $e_{ij}^{(1)}$ and $e_{ij}^{(2)}$ refer to deformations of the body and tensor $e_{ij}^{(3)}$ represents a pure rotation. For the purpose of the analysis we may disregard the last tensor, $e_{ij}^{(3)}$, and combine the remaining tensors into one

$$\varepsilon_{ij} = e_{ij}^{(1)} + e_{ij}^{(2)} = \begin{pmatrix} \dfrac{\partial u}{\partial x} & \dfrac{1}{2}\left(\dfrac{\partial u}{\partial y} + \dfrac{\partial v}{\partial x}\right) & \dfrac{1}{2}\left(\dfrac{\partial u}{\partial z} + \dfrac{\partial w}{\partial x}\right) \\[2mm] \dfrac{1}{2}\left(\dfrac{\partial v}{\partial x} + \dfrac{\partial u}{\partial y}\right) & \dfrac{\partial v}{\partial y} & \dfrac{1}{2}\left(\dfrac{\partial v}{\partial z} + \dfrac{\partial w}{\partial y}\right) \\[2mm] \dfrac{1}{2}\left(\dfrac{\partial w}{\partial x} + \dfrac{\partial u}{\partial z}\right) & \dfrac{1}{2}\left(\dfrac{\partial w}{\partial y} + \dfrac{\partial v}{\partial z}\right) & \dfrac{\partial w}{\partial z} \end{pmatrix} \qquad [2.29]$$

The tensor of Equation (2.29) is symmetric. Introducing the following expressions

$$\varepsilon_x = \frac{\partial u}{\partial x}, \qquad \varepsilon_y = \frac{\partial v}{\partial y}, \qquad \varepsilon_z = \frac{\partial w}{\partial z} \qquad [2.30]$$

and

$$\begin{aligned} \gamma_{xy} &= \frac{\partial u}{\partial y} + \frac{\partial v}{\partial x}, & \gamma_{yx} &= \frac{\partial v}{\partial x} + \frac{\partial u}{\partial y}, \\[2mm] \gamma_{yz} &= \frac{\partial v}{\partial z} + \frac{\partial w}{\partial y}, & \gamma_{zy} &= \frac{\partial w}{\partial y} + \frac{\partial v}{\partial z}, \\[2mm] \gamma_{zx} &= \frac{\partial w}{\partial x} + \frac{\partial u}{\partial z}, & \gamma_{xz} &= \frac{\partial u}{\partial z} + \frac{\partial w}{\partial x}. \end{aligned} \qquad [2.30']$$

tensor ε_{ij} becomes

$$\varepsilon_{ij} = \begin{pmatrix} \varepsilon_x & \dfrac{\gamma_{xy}}{2} & \dfrac{\gamma_{xz}}{2} \\[2mm] \dfrac{\gamma_{yx}}{2} & \varepsilon_y & \dfrac{\gamma_{yz}}{2} \\[2mm] \dfrac{\gamma_{zx}}{2} & \dfrac{\gamma_{zy}}{2} & \varepsilon_z \end{pmatrix} \qquad [2.31]$$

We thus defined the body deformation in mathematical terms.

2.1.3 Stress-strain relation

In Sections 2.1.1 and 2.1.2 we investigated respectively the stress and strain conditions at a point under loading. Now let us consider the mutual relationship of the two in elastic state. Experimentally it is proven that for isotropic and linearly elastic materials the following correlation between the strains and the stresses applies

[2.32]

$$\varepsilon_x = \frac{1}{E}\Big[\sigma_x - \nu\big(\sigma_y + \sigma_z\big)\Big]$$

$$\varepsilon_y = \frac{1}{E}\Big[\sigma_y - \nu\big(\sigma_z + \sigma_x\big)\Big]$$

$$\varepsilon_z = \frac{1}{E}\Big[\sigma_z - \nu\big(\sigma_x + \sigma_y\big)\Big]$$

and

[2.32']

$$\gamma_{xy} = \frac{2(1+\nu)}{E}\tau_{xy}$$

$$\gamma_{yz} = \frac{2(1+\nu)}{E}\tau_{yz}$$

$$\gamma_{zx} = \frac{2(1+\nu)}{E}\tau_{zx}$$

where E and ν are experimentally determined constants. E is the elasticity modulus of Young and ν is Poisson ratio. The correlation between stresses and strains expressed by Equations (2.32) and (2.32') is known as Hooke's law. Equations (2.32') are sometimes written in abbreviated form

[2.32'']

$$\gamma_{xy} = \frac{\tau_{xy}}{G}, \quad \gamma_{yz} = \frac{\tau_{yz}}{G}, \quad \gamma_{zx} = \frac{\tau_{zx}}{G}$$

where G is the shear modulus defined by

[2.33]

$$G = \frac{E}{2(1+\nu)}$$

2.1.4 Energy and work

We turn now to the analysis of work performed by stresses $\sigma_x, \sigma_y, \sigma_z, \tau_{xy}, \tau_{yz}$ and τ_{zx} during deformation within the loaded body. The work is transferred into an elastic strain energy of the body.

Elastic strain energy

Consider deformation work performed by stress σ_x along coordinate x, as illustrated by the elemental cube shown in Figure 2.1. The stress rises from 0 to σ_x, causing a change in length of the cube equal to $(\partial u/\partial x)\,dx$. The work performed by stress σ_x equals

$$\frac{1}{2}\left(\sigma_x \cdot dy \cdot dz\right)\frac{\partial u}{\partial x}dx = \frac{1}{2}\sigma_x \varepsilon_x dx \cdot dy \cdot dz \qquad [2.34]$$

Similarly, work performed by shear stresses τ_{xy} and τ_{yx} is

$$\frac{1}{2}\left(\tau_{xy}dy \cdot dz\right)\left(\frac{\partial u}{\partial y}+\frac{\partial v}{\partial x}\right)dx + \frac{1}{2}\left(\tau_{yx}dz \cdot dx\right)\left(\frac{\partial v}{\partial x}+\frac{\partial u}{\partial y}\right)dy \qquad [2.35]$$

$$= \frac{1}{2}\tau_{xy}\gamma_{xy}dx \cdot dy \cdot dz$$

The total deformation work, performed by all stresses, equals

$$dW_i = W_{io} \cdot dx \cdot dy \cdot dz \qquad [2.36]$$

$$= \frac{1}{2}\left(\sigma_x \varepsilon_x + \sigma_y \varepsilon_y + \sigma_z \varepsilon_z + \tau_{xy}\gamma_{xy} + \tau_{yz}\gamma_{yz} + \tau_{zx}\gamma_{zx}\right)dx \cdot dy \cdot dz$$

To reflect the strain energy density, we express W_{io}, the total work per unit volume, as

$$W_{io} = \frac{1}{2}\left(\sigma_x \varepsilon_x + \sigma_y \varepsilon_y + \sigma_z \varepsilon_z + \tau_{xy}\gamma_{xy} + \tau_{yz}\gamma_{yz} + \tau_{zx}\gamma_{zx}\right) \qquad [2.37]$$

Using principal strains, ε_1, ε_2 and ε_3, and assuming that the principal strain directions are the same as the principal stress directions,[1] Equation (2.37) takes the form

$$W_{io} = \frac{1}{2}\left(\sigma_1 \varepsilon_1 + \sigma_2 \varepsilon_2 + \sigma_3 \varepsilon_3\right) \qquad [2.38]$$

Based on Hooke's correlation, the latter becomes

$$W_{io} = \frac{1}{2}\left(\sigma_1^2 + \sigma_2^2 + \sigma_3^2\right) - \frac{\nu}{E}\left(\sigma_1\sigma_2 + \sigma_2\sigma_3 + \sigma_3\sigma_1\right) \qquad [2.39]$$

We thus derived the expression of strain energy density as a function of principal stresses.

Virtual work

In application of the elasticity theory the principle of virtual work is of special significance for numerical derivation of stresses and strains. A detailed presentation of the principle follows.

Consider a loaded body in equilibrium with corresponding displacements and strains, {a} and {ε}, respectively. Let us assume that in addition each point of the body undergoes, within prescribed constraints, a small reversible displacement. It is called virtual displacement and is denoted in a vector form by

$$\{\delta \mathbf{u}\} = \begin{Bmatrix} \delta u \\ \delta v \\ \delta w \end{Bmatrix}$$

[2.40]

Differentiating the above equation, we obtain the virtual strains

[2.41]

$$\delta \varepsilon_x = \frac{\partial}{\partial x} \delta u$$

$$\delta \varepsilon_y = \frac{\partial}{\partial y} \delta v$$

$$\delta \varepsilon_z = \frac{\partial}{\partial z} \delta w$$

$$\delta \gamma_{xy} = \frac{\partial}{\partial y} \delta u + \frac{\partial}{\partial x} \delta v$$

$$\delta \gamma_{yz} = \frac{\partial}{\partial z} \delta v + \frac{\partial}{\partial y} \delta w$$

$$\delta \gamma_{zx} = \frac{\partial}{\partial x} \delta w + \frac{\partial}{\partial z} \delta u$$

The work performed by stresses along the virtual strains equals

[2.42]

$$\delta W_i = \iiint_{vol} \begin{pmatrix} \dfrac{\partial W_{io}}{\partial \varepsilon_x} \delta \varepsilon_x + \dfrac{\partial W_{io}}{\partial \varepsilon_y} \delta \varepsilon_y + \dfrac{\partial W_{io}}{\partial \varepsilon_z} \delta \varepsilon_z \\ + \dfrac{\partial W_{io}}{\partial \varepsilon_{xy}} \delta \gamma_{xy} + \dfrac{\partial W_{io}}{\partial \varepsilon_{yz}} \delta \gamma_{yz} + \dfrac{\partial W_{io}}{\partial \varepsilon_{zx}} \delta \gamma_{zx} \end{pmatrix} dx \cdot dy \cdot dz$$

Therefore the virtual internal work can be expressed as

[2.43]

$$\delta W_i = \iiint_{vol} \begin{pmatrix} \sigma_x \delta \varepsilon_x + \sigma_y \delta \varepsilon_y + \sigma_z \delta \varepsilon_z \\ + \tau_{xy} \delta \gamma_{xy} + \tau_{yz} \delta \gamma_{yz} + \tau_{zx} \delta \gamma_{zx} \end{pmatrix} dx \cdot dy \cdot dz$$

On the other hand, the virtual work of applied external forces equals

$$\delta W_F = \iiint_{vol} \left(F_x \delta u + F_y \delta v + F_z \delta w \right) dx \cdot dy \cdot dz + \iint_{surf} \left(p_x \delta u + p_y \delta v + p_z \delta w \right) dS \quad [2.44]$$

where F_i denotes the applied body forces and p_i the applied surface pressures.[2] In equilibrium the sum of Equations (2.43) and (2.44) must equal zero

$$-\delta W_i + \delta W_F = 0 \qquad [2.45]$$

The above mathematical form expresses the principle of virtual work which states: *In a body in equilibrium the total work, performed by internal stresses and applied external forces, must result in zero for all virtual displacements and virtual strains.*

2.1.5 Equivalent stresses and strains

Working machine parts in general are subject to multiaxial loading which results in multiaxial stresses and strains. To utilize for design purposes the data derived from one-dimensional experiments, it is necessary to replace the multiaxial stresses or strains with equivalent one-dimensional stresses or strains which we denote as σ_e and σ_e respectively.

There are several known theories. used as criteria, to derive equivalent stresses and strains.[3] Those presented here were chosen for their special applicability to fatigue analysis.

von Mises criteria

von Mises theory was originally used to derive equivalent stresses. It was subsequently applied in defining equivalent strains as well.

Equivalent stress The tensor of equivalent stress σ_e. equals

$$\sigma_{ij} = \begin{pmatrix} \sigma_e & 0 & 0 \\ 0 & 0 & 0 \\ 0 & 0 & 0 \end{pmatrix} \qquad [2.46]$$

von Mises theory assumes that hydrostatic pressure or tension does not affect the critical behavior of a stressed body. In accordance with this assumption, we shall consider the stress deviator tensor instead. It follows that

$$\begin{pmatrix} \sigma_e - \dfrac{\sigma_e}{3} & 0 & 0 \\ 0 & -\dfrac{\sigma_e}{3} & 0 \\ 0 & 0 & -\dfrac{\sigma_e}{3} \end{pmatrix} = \begin{pmatrix} \dfrac{2\sigma_1 - \sigma_2 - \sigma_3}{3} & 0 & 0 \\ 0 & \dfrac{2\sigma_2 - \sigma_3 - \sigma_1}{3} & 0 \\ 0 & 0 & \dfrac{2\sigma_3 - \sigma_1 - \sigma_2}{3} \end{pmatrix} \qquad [2.47]$$

Let us compare the second invariant of the deviator tensor of the equivalent stress with that of the of the multiaxial stresses, as per Equation (2.20). We obtain

$$\frac{1}{3}\sigma_e^2 = \frac{1}{6}\left[\left(\sigma_1 - \sigma_2\right)^2 + \left(\sigma_2 - \sigma_3\right)^2 + \left(\sigma_3 - \sigma_1\right)^2 \right] \qquad [2.48]$$

whereby the equivalent stress becomes

[2.49]

$$\sigma_e = \sqrt{\frac{1}{2}\left[\left(\sigma_1-\sigma_2\right)^2+\left(\sigma_2-\sigma_3\right)^2+\left(\sigma_3-\sigma_1\right)^2\right]}$$

$$= \sqrt{\frac{1}{2}\left[\left(\sigma_x-\sigma_y\right)^2+\left(\sigma_y-\sigma_z\right)^2+\left(\sigma_z-\sigma_x\right)^2+6\left(\tau_{xy}^2+\tau_{yz}^2+\tau_{zx}^2\right)\right]}$$

Equation (2.48) is known as von Mises criterion. Sometimes the above equation is referred to as criteria for the distortion energy theory (derived by M. I. Huber and H. Hencky), as well as octahedral theory. This is because it can also be obtained from these respective theories. See Reference 3.

   Equivalent strain   The equivalent strain is used, similarly to equivalent stress, to correlate three-dimensional strains. According to Hooke's law, the one-dimensional equivalent stress condition provides three principal strain components

[2.50]

$$\varepsilon_1 = \frac{1}{E}\sigma_e = \varepsilon_e$$

$$\varepsilon_2 = -\frac{1}{E}v\sigma_e = -v\varepsilon_e$$

$$\varepsilon_3 = -\frac{1}{E}v\sigma_e = -v\varepsilon_e$$

Consequently we derive the equivalent strain tensor

[2.51]

$$\varepsilon_{ij} = \begin{pmatrix} \varepsilon_e & 0 & 0 \\ 0 & -v\varepsilon_e & 0 \\ 0 & 0 & -v\varepsilon_e \end{pmatrix}$$

Repeating the procedure used above for stress deviator tensors and comparing the strain deviator tensors of the one- and three-dimensional cases, we get

[2.52]

$$\begin{pmatrix} \varepsilon_e - \dfrac{\varepsilon_e - 2v\varepsilon_e}{3} & 0 & 0 \\ 0 & -v\varepsilon_e - \dfrac{\varepsilon_e - 2v\varepsilon_e}{3} & 0 \\ 0 & 0 & -v\varepsilon_e - \dfrac{\varepsilon_e - 2v\varepsilon_e}{3} \end{pmatrix}$$

$$= \begin{pmatrix} \dfrac{2\varepsilon_1 - \varepsilon_2 - \varepsilon_3}{3} & 0 & 0 \\ 0 & \dfrac{2\varepsilon_2 - \varepsilon_3 - \varepsilon_1}{3} & 0 \\ 0 & 0 & \dfrac{2\varepsilon_3 - \varepsilon_1 - \varepsilon_2}{3} \end{pmatrix}$$

From this one obtains von Mises equivalent strain criterion

$$\varepsilon_e = \frac{1}{(1+\nu)\sqrt{2}}\sqrt{(\varepsilon_1-\varepsilon_2)^2+(\varepsilon_2-\varepsilon_3)^2=(\varepsilon_3-\varepsilon_1)^2} \qquad [2.53]$$

Tresca criteria

A different approach was used by Tresca who introduced the equivalent shear stresses. It is also true that only later his theory was extended to include equivalent shear strains.

Equivalent shear stress   The criterion correlates the equivalent shear stress with three-dimensional stresses. (See Section 2.1.1). According to this, the equivalent shear stress equals the largest of the principal shear stresses as follows,

$$\tau_e = \frac{|\sigma_1-\sigma_2|}{2} \qquad [2.54]$$

$$\tau_e = \frac{|\sigma_2-\sigma_3|}{2}$$

$$\tau_e = \frac{|\sigma_3-\sigma_1|}{2}$$

The principal stress differences are independent of mean stress σ_m, see Equation (2.18), and can be expressed as

$$\sigma_1-\sigma_2 = s_1-s_2 \qquad [2.55]$$
$$\sigma_2-\sigma_3 = s_2-s_3$$
$$\sigma_3-\sigma_1 = s_3-s_1$$

It follows that Tresca shear stress criterion is a function of deviatoric stresses only.

Stress intensity   A modified form of the Tresca stress criterion introduces an equivalent normal stress, called stress intensity, which is the largest of the following three expressions

$$\sigma_e = |\sigma_1-\sigma_2| \qquad [2.56]$$

$$\sigma_e = |\sigma_2-\sigma_3|$$

$$\sigma_e = |\sigma_3-\sigma_1|$$

The stress intensity is used by the ASME code for construction of pressure vessels.[4]

Equivalent shear strain   This criterion correlates an equivalent shear strain with three-dimensional strains. Using the tensor of the equivalent strain, Equation (2.51), as a starting point, we obtain the expression of the principal shear strain

$$\gamma_e = \frac{|\varepsilon_1 - \varepsilon_2|}{2} = \frac{1+\nu}{2}\varepsilon_e \qquad [2.57]$$

Consequently, the equivalent shear strain equals the largest of the following three expressions

$$\gamma_e = \frac{1+\nu}{2}\varepsilon_e = \frac{|\varepsilon_1 - \varepsilon_2|}{2} \qquad [2.58]$$

$$\gamma_e = \frac{1+\nu}{2}\varepsilon_e = \frac{|\varepsilon_2 - \varepsilon_3|}{2}$$

$$\gamma_e = \frac{1+\nu}{2}\varepsilon_e = \frac{|\varepsilon_3 - \varepsilon_1|}{2}$$

Maximum principal stress and strain criteria

The simplest approach is to correlate the equivalent stress or strain with the respective maximum principal values. The maximum principal stress criterion is expressed by the equation

$$\sigma_e = \sigma_1 \qquad [2.59]$$

where σ_1 is the maximum principal stress. Similarly, the maximum principal strain criterion is expressed as follows

$$\varepsilon_e = \varepsilon_1 \qquad [2.60]$$

where ε_1 is the maximum principal strain.

2.2 PLASTICITY

A machine part under high loading may reach a condition where the stresses pass the yield point and the body enters a plastic state. To analyze the stresses and strains occurring during this condition, the theory of plasticity applies. This section includes the discussion of multiaxial stresses, stress-stress relations and the underlying theories applicable to the plastic state.

2.2.1 Yield function

To predict the behavior of plastic deformation for a multiaxial stress condition, a mathematical function, called yield function, is applied. Its purpose is to correlate a one-dimensional equivalent stress with multiaxial stresses in plastic domain. To derive the most commonly used yield functions in machine design, let us first look at the one applied to isotropic hardening materials (explained further below). It is expressed as

$$F(\sigma_{ij}) = f(\sigma_{ij}) - Y(\kappa) = 0 \tag{2.61}$$

where $f(\sigma_{ij})$ expresses a yield criterion and $Y(\kappa)$ is a measure of plastic deformation in terms of parameter κ. During plastic deformation parameter κ increases, reflecting the stress-strain curve above the yield point. See Figure 2.5 which presents typical experimental data for a one-dimensional specimen.

One of the yield functions based on von Mises theory (see Section 2.1.5) takes a form of

$$F(\sigma_{ij}) = \frac{1}{6}\left[(\sigma_1 - \sigma_2) + (\sigma_2 - \sigma_3) + (\sigma_3 - \sigma_1)\right] - \frac{1}{3}\sigma_e^2 = 0 \tag{2.62}$$

with the yield criterion

$$f(\sigma_{ij}) = \frac{1}{6}\left[(\sigma_1 - \sigma_2) + (\sigma_2 - \sigma_3) + (\sigma_3 - \sigma_1)\right] \tag{2.63}$$

and the plastic deformation function

$$Y(\sigma_e) = \frac{1}{3}\sigma_e^2 \tag{2.64}$$

σ_e denotes a one-dimensional equivalent stress above the yield point.

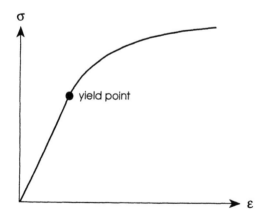

Figure 2.5 Stress-strain relation of a one-dimensional specimen. The yield point indicates the onset of plastic deformation.

Another important yield function used in machine design is based on Tresca theory

$$F\left(\sigma_{ij}\right) = \frac{\sigma_1 - \sigma_3}{2} - \frac{1}{2}\sigma_e^2 = 0 \qquad [2.65]$$

with the yield criterion

$$f\left(\sigma_{ij}\right) = \frac{\sigma_1 - \sigma_3}{2} \qquad [2.66]$$

and the plastic deformation function

$$Y\left(\sigma_e\right) = \frac{1}{2}\sigma_e \qquad [2.67]$$

We can visualize yield function $F(\sigma_{ij})$ as a three-dimensional yield surface in a stress space defined by coordinates $(\sigma_1, \sigma_2, \sigma_3)$.[5-7] Figure 2.6 shows projections of von Mises and Tresca yield functions in the stress space.

In consideration of differences in plastic deformation due to different types of materials, the behavior of yield surface is what determines the yield function. The behavior of yield surfaces is classified as follows

(a) Isotropic hardening where the yield surface during loading expands in the stress space equally in all directions (*i*); the yield function is expressed by Equation (2.61) above.

(b) Kinematic hardening where the yield surface in some directions expands, while in others it contracts (see Bauschinger effect, Reference 7). Here the yield function is

$$F\left(\sigma_{ij}\right) = f\left(\sigma_{ij} - \alpha_{ij}\right) - Y\left(\kappa\right) = 0 \qquad [2.68]$$

where α_{ij} denotes a translation of the yield surface.

(c) No hardening in perfectly plastic material where the yield surface remains constant and the slope of stress-strain line, is zero in plastic domain. The corresponding yield function becomes

$$F\left(\sigma_{ij}\right) = f\left(\sigma_{ij}\right) - \text{const} = 0 \qquad [2.69]$$

For fatigue analysis the isotropic hardening presents the most practical application. The use of kinematic hardening is impractical because the Bauschinger effect tends to disappear after a large number of fluctuations. In fact, alternative (c), perfect plasticity is an assumption which is not always applicable to fatigue analysis because it may introduce inaccuracies. For other yield functions, less often applied in fatigue analysis, see References 5 and 6.

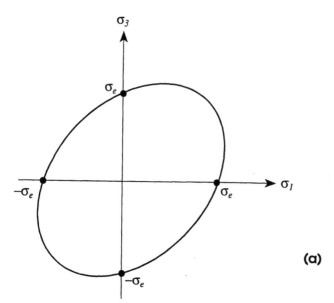

Figure 2.6 Projections of yield surfaces in (σ_1, σ_3) plane: (a) von Mises yield surface; (b) Tresca yield surface.

(a)

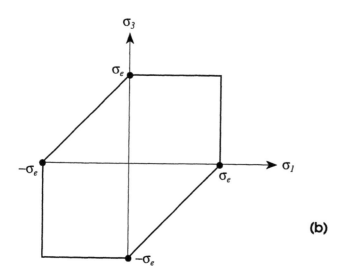

(b)

2.2.2 Stress-strain relation

The theory of stress-strain relation applicable to plastic state is based on three hypotheses: first, that hydrostatic pressure or tension does not affect plastic deformation; second, that material density is not affected by purely plastic deformation; and third, that plastic strain increments act in direction normal to the yield surface.

The first hypothesis reflects the fact that only deviatoric stresses are considered. The second hypothesis is expressed by the equation

$$d\varepsilon_1^p + d\varepsilon_2^p + d\varepsilon_3^p = 0 \qquad [2.70]$$

which means that the material is incompressible with respect to plastic strains. The third hypothesis, known as the flow rule, can be expressed in the form

$$d\varepsilon_{ij}^p = d\lambda \frac{\partial f}{\partial \sigma_{ij}} \qquad [2.71]$$

where $d\lambda$ is a nonnegative proportionality factor. The derivative on the right side of the equation represents a gradient of the yield criterion.

Associated flow rule

The flow rule forms the mathematical basis for correlating stresses and strains in plastic domain. It views the process of plastic deformation as plastic strain flow whose direction is normal to the yield surface. Since there are a number of theoretical criteria, the resulting correlations of stresses and strains, based on a particular criterion, are different. This is the explanation for the all encompassing term of associated flow rule.

For instance, defining the gradient of yield criterion $f(\sigma_{ij})$ per von Mises theory, the flow rule takes the form

$$d\varepsilon_1^p = \frac{2}{3} d\lambda \left[\sigma_1 - \frac{1}{2}(\sigma_2 + \sigma_3) \right] = d\lambda\, s_1 \qquad [2.72]$$

$$d\varepsilon_2^p = \frac{2}{3} d\lambda \left[\sigma_2 - \frac{1}{2}(\sigma_3 + \sigma_1) \right] = d\lambda\, s_2$$

$$d\varepsilon_3^p = \frac{2}{3} d\lambda \left[\sigma_3 - \frac{1}{2}(\sigma_1 + \sigma_2) \right] = d\lambda\, s_3$$

while per Tresca, the flow rule becomes

$$\partial \varepsilon_1^p = \frac{1}{2} d\lambda \qquad [2.73]$$

$$\partial \varepsilon_2^p = 0$$

$$\partial \varepsilon_3^p = -\frac{1}{2} d\lambda$$

Stress-strain correlations

Let us review the application of the associated flow rule, first, according to von Mises yield function and, then according to Tresca.

von Mises theory   According to von Mises, the equivalent stress equals

$$\sigma_{e,M} = \sqrt{\frac{1}{2}\left[\left(\sigma_1-\sigma_2\right)^2 + \left(\sigma_2-\sigma_3\right)^2 + \left(\sigma_3-\sigma_1\right)^2\right]}$$

$$= \sqrt{\frac{3}{2}\left(s_1^2 + s_2^2 + s_3^2\right)}$$

[2.74]

By analogy, one can define an equivalent plastic strain increment

$$d\varepsilon_{p,M} = \frac{\sqrt{2}}{3}\sqrt{\left(d\varepsilon_1^p - d\varepsilon_2^p\right)^2 + \left(d\varepsilon_2^p - d\varepsilon_3^p\right)^2 + \left(d\varepsilon_3^p - d\varepsilon_1^p\right)^2}$$

$$= \sqrt{\frac{2}{3}\left[\left(d\varepsilon_1^p\right)^2 + \left(d\varepsilon_2^p\right)^2 + \left(d\varepsilon_3^p\right)^2\right]}$$

[2.75]

The latter equation is derived from Equation (2.53), assuming a Poisson ratio for plastic material $\nu = 0.5$. Consequently, one obtains the proportionality factor, $d\lambda$, in the form

$$d\lambda = \frac{3}{2}\frac{d\varepsilon_{p,M}}{\sigma_{e,M}}$$

[2.76]

Substituting $d\lambda$ in Equation (2.72), the final stress-strain correlations become

$$d\varepsilon_1^p = \frac{3}{2}\frac{d\varepsilon_{p,M}}{\sigma_{e,M}}s_1$$

[2.77]

$$d\varepsilon_2^p = \frac{3}{2}\frac{d\varepsilon_{p,M}}{\sigma_{e,M}}s_2$$

$$d\varepsilon_3^p = \frac{3}{2}\frac{d\varepsilon_{p,M}}{\sigma_{e,M}}s_3$$

Tresca yield theory   The equivalent plastic strain increment, according to Tresca criterion, is based on Equation (2.58) where the plastic strain components are substituted by the respective increments. The Poisson ratio $\nu = 0.5$ is applied. Consequently,

$$d\varepsilon_{p,T} = \frac{2}{3}\left(d\varepsilon_1^p - d\varepsilon_3^p\right)$$

[2.78]

On the other hand, the equivalent stress equals

$$\sigma_{e,T} = \sigma_1 - \sigma_3 = s_1 - s_3$$

[2.79]

It follows from the above that

$$\frac{d\varepsilon_1^p - d\varepsilon_3^p}{s_1 - s_3} = \frac{3}{2}\frac{d\varepsilon_{p,T}}{\sigma_{e,T}}$$ [2.80]

The final Tresca stress-strain correlations in 1- and 3- direction become

$$d\varepsilon_1^p = \frac{3}{2}\frac{d\varepsilon_{p,T}}{\sigma_{e,T}}s_1$$ [2.81]

$$d\varepsilon_3^p = \frac{3}{2}\frac{d\varepsilon_{p,T}}{\sigma_{e,T}}s_3$$

The plastic strain increment in 2-direction is

$$d\varepsilon_2^p = 0$$ [2.82]

In accordance with Tresca, this confirms the fact that s_2 does not affect the yielding.

Prandtl-Reuss equations   Historically original stress-strain correlations were based on experimental observations that showed that plastic strain increments are proportional to the instantaneous stress deviations and act in the same direction. The corresponding mathematical expression had the form

$$\frac{d\varepsilon_x^p}{s_x} = \frac{d\varepsilon_y^p}{s_y} = \frac{d\varepsilon_z^p}{s_z} = \frac{d\gamma_{yz}^p}{2\tau_{yz}} = \frac{d\gamma_{zx}^p}{2\tau_{zx}} = \frac{d\gamma_{xy}^p}{2\tau_{xy}}$$ [2.83]

known as the Prandtl-Reuss equations. These theories have confirmed the observations.

2.2.3 Work hardening

Work hardening, or strain hardening, refers to decreasing plastic flow due to increased material resistance with greater plastic deformation. It is characterized by the growth of function $Y(\kappa)$ in Equation (2.61). (See Section 2.2.1.) To facilitate the use of one-dimensional experimental data for numerical application, the function Y is expressed here in terms of plastic work.

Plastic work

We start with a mathematical definition of work in plastic domain. The total work equals

$$dW = \sigma_{ij}d\varepsilon_{ij} = \sigma_{ij}\left(d\varepsilon_{ij}^e + d\varepsilon_{ij}^p\right)$$ [2.84]

Part of the work which pertains to the plastic strain increment (plastic work) equals

$$dW^p = \sigma_{ij}d\varepsilon_{ij}^p \qquad [2.85]$$

The latter is equivalent to

$$dW^p = \sigma_e d\varepsilon_p \qquad [2.86]$$

(See Reference 5.) We again refer to Equation (2.61) and apply it to plastic work, substituting parameter κ with W_p

$$f\left(\sigma_{ij}\right) - Y\left(W^p\right) = 0 \qquad [2.87]$$

By analogy with a one-dimensional stress-strain correlation, the equivalent stress is can presented as a function of the effective plastic strain

$$\sigma_e = H\left(\varepsilon_p\right) \qquad [2.88]$$

It follows from Equations (2.85) and (2.87) that

$$f\left(\sigma_{ij}\right) - Y\left(H\right) = 0 \qquad [2.89]$$

The latter equation lets us define the stress-strain correlation in terms of experimental data. Thus, using von Mises theory, it can be expressed as

$$d\varepsilon_{ij}^p = \frac{3}{2}\frac{d\sigma_e}{H'\sigma_e}s_{ij} \qquad [2.90]$$

where H' is the slope of one-dimensional stress-strain curve obtained from tests,

$$H' = \frac{d\sigma_e}{d\varepsilon_p} \qquad [2.91]$$

2.2.4 Incremental and deformation theories

The associated flow rules in plastic deformation and the derived stress-strain relations refer to the incremental theory of plasticity as presented above. The theory of plasticity also includes another approach called deformation theory. The reasoning of the deformation theory is different from that of incremental theory. However, the same associated flow rules and the derived stress-strain correlations are used. The difference is that the respective mathematical expressions are presented in an integral form. (See References 6 and 7.) The deformation theory is based on a total approach, independent of the loading path. In the analysis of stresses and strains in general, whenever the stress components rise in nonequal ratios, non-proportional loading, the incremental theory is applied. On the other hand, when proportional loading takes place the deformation theory may also be used.

We refer to stress-strain correlations based on von Mises and Tresca. The mathematical expressions, for convenience, are taken from von Mises theory. The conclusions, however, are equally applicable to both theories. The stress-strain correlation used in the incremental theory equals

$$d\varepsilon_{ij}^p = \frac{3}{2}\frac{s_{ij}}{\sigma_e}d\varepsilon_p$$

[2.92]

and in the deformation theory it takes the form

$$\varepsilon_{ij}^p = \frac{3}{2}\frac{s_{ij}}{\sigma_e}\varepsilon_p$$

[2.93]

The numerical analysis of engineering problems is predominantly based on the incremental theory. The deformation theory is applicable when the strain is independent of the loading path in the stress space. When strains caused by stresses along respective paths are different, the problem becomes path dependent and one must use the incremental theory.

2.3
FINITE ELEMENT METHOD

The finite element (FE) method, as used in numerical analysis, considers a loaded body as an assembly of discrete building blocks. The blocks (elements) are considered as severed from each other and are joined only at specific points (nodes), forming a network. The number of elements is determined by the capabilities of the computer and the desired accuracy of the results. We assume the existence of forces within the assembly, acting upon the elements through interconnecting nodes. The stress analysis pertains to computing displacements at the nodes and using the results in determination of strains and stresses.

The following two compatibility conditions must exist to provide the mathematical justification for the stress and strain distribution in the elements must exist:

(a) elements joined by a particular node have at the respective node the same displacement;
(b) the interface between two elements is subject to displacements that are functions of the corresponding nodes on the interface, and those nodes only.

2.3.1 The equilibrium condition

The equilibrium condition in a FE model is expressed by a balance of nodal forces and stresses in the elements. The stresses are derived assuming a polynomial displacement distribution using shape functions.[8,9]

Shape function

To derive the stress and strain distribution in the elements we must determine a polynomial relationship called the shape function. An illustrative example is used for the derivation of the shape function. Let us consider the FE model shown in Figure 2.7. We select element *e* and denote its nodes by 1 to 8 in Figure 2.7(b). The displacement of any point within element *e* is a function of location and can be expressed by

$$u = N_i u_i$$
$$v = N_i v_i \qquad\qquad\qquad\qquad [2.94]$$
$$w = N_i w_i$$

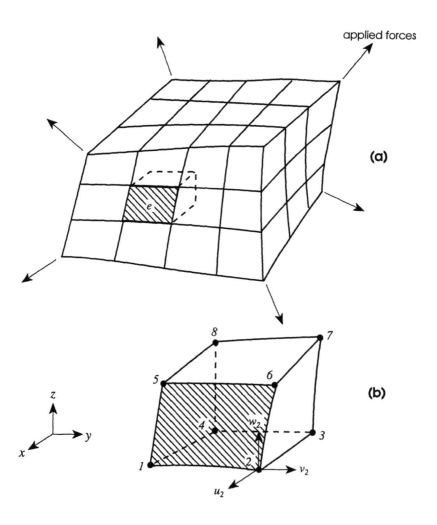

Figure 2.7 A finite element model; (a) assembly of elements representing a body under load; (b) single element.

The polynomials N_i are given in terms of coordinates (x,y,z) and u_i, v_i and w_i denote displacements at the nodes. In vectorial form, Equations (2.94) becomes

$$\begin{Bmatrix} u \\ v \\ w \end{Bmatrix} = \begin{bmatrix} N_1 & 0 & 0 & N_2 & 0 & 0 & \cdots & N_i & 0 & 0 \\ 0 & N_1 & 0 & 0 & N_2 & 0 & 0 & \cdots & N_i & 0 \\ 0 & 0 & N_1 & 0 & 0 & N_2 & 0 & 0 & \cdots & N_i \end{bmatrix} \begin{Bmatrix} u_1 \\ v_1 \\ w_1 \\ u_2 \\ v_2 \\ w_2 \\ \cdot \\ \cdot \\ \cdot \\ u_i \\ v_i \\ w_i \end{Bmatrix} \qquad [2.95]$$

or in the matrix form

$$\{u\} = [N]\{u_e\} \qquad [2.96]$$

Matrix [N] shown above is designated as a shape function.

Strains and stresses

To define the strains we use Equations (2.30) and (2.30′) of Section 2.1. In view of the definition of displacements, as per Equations (2.94), the strains equal

$$[2.97]$$

$$\varepsilon_x = \frac{\partial}{\partial x} N_i u_i$$

$$\varepsilon_y = \frac{\partial}{\partial y} N_i v_i$$

$$\varepsilon_z = \frac{\partial}{\partial z} N_i w_i$$

$$\gamma_{xy} = \frac{\partial}{\partial x} N_i v_i + \frac{\partial}{\partial y} N_i u_i$$

$$\gamma_{yz} = \frac{\partial}{\partial y} N_i w_i + \frac{\partial}{\partial z} N_i v_i$$

$$\gamma_{zx} = \frac{\partial}{\partial z} N_i u_i + \frac{\partial}{\partial x} N_i w_i$$

The strains can be expressed as

$$\begin{Bmatrix} \varepsilon_x \\ \varepsilon_y \\ \varepsilon_z \\ \gamma_{xy} \\ \gamma_{yz} \\ \gamma_{zx} \end{Bmatrix} = \begin{Bmatrix} \dfrac{\partial}{\partial x} \\[2mm] \dfrac{\partial}{\partial y} \\[2mm] \dfrac{\partial}{\partial z} \\[2mm] \dfrac{\partial}{\partial x} + \dfrac{\partial}{\partial y} \\[2mm] \dfrac{\partial}{\partial y} + \dfrac{\partial}{\partial z} \\[2mm] \dfrac{\partial}{\partial z} + \dfrac{\partial}{\partial x} \end{Bmatrix} \begin{bmatrix} \mathbf{N} \end{bmatrix} \{\mathbf{u}_e\}$$

[2.98]

which becomes

$$\{\boldsymbol{\varepsilon}\} = \begin{bmatrix} \mathbf{B} \end{bmatrix} \{\mathbf{u}_e\}$$

[2.99]

Matrix [B] is the result of a differential operation upon [N].

In order to determine the stresses, we use Hooke's law of elasticity, as per Equations (2.32) and (2.32′), in a matrix form

$$\{\boldsymbol{\varepsilon}\} = \begin{bmatrix} \mathbf{D} \end{bmatrix}^{-1} \{\boldsymbol{\sigma}\}$$

[2.100]

The stress vector is expressed by the reciprocal matrix equation

$$\{\boldsymbol{\sigma}\} = \begin{bmatrix} \mathbf{D} \end{bmatrix} \{\boldsymbol{\varepsilon}\} = \begin{bmatrix} \mathbf{D} \end{bmatrix} \begin{bmatrix} \mathbf{B} \end{bmatrix} \{\mathbf{u}_e\}$$

[2.101]

where matrix [D] is derived from Equations (2.32) and (2.32′). It is called the elasticity matrix and equals

$$\begin{bmatrix} \mathbf{D} \end{bmatrix} = \frac{E}{1+\nu} \begin{bmatrix} \dfrac{1-\nu}{1-2\nu} & \dfrac{\nu}{1-2\nu} & \dfrac{\nu}{1-2\nu} & 0 & 0 & 0 \\[3mm] \dfrac{\nu}{1-2\nu} & \dfrac{1-\nu}{1-2\nu} & \dfrac{\nu}{1-2\nu} & 0 & 0 & 0 \\[3mm] \dfrac{\nu}{1-2\nu} & \dfrac{\nu}{1-2\nu} & \dfrac{1-\nu}{1-2\nu} & 0 & 0 & 0 \\[3mm] 0 & 0 & 0 & \dfrac{1}{2} & 0 & 0 \\[3mm] 0 & 0 & 0 & 0 & \dfrac{1}{2} & 0 \\[3mm] 0 & 0 & 0 & 0 & 0 & \dfrac{1}{2} \end{bmatrix}$$

[2.102]

Equations (2.99) and (2.101) are used by the FE method to compute stresses and strains from node displacements.

Matrix [D], as expressed by Equation (2.102), applies to three-dimensional bodies. For two-dimensional bodies it is reduced to a three-by-three matrix as shown below in a brief presentation of two cases, plane stress and plane strain. (For a detailed derivation of the [D] matrices the reader is directed to References 8 and 9.)

Plane stress Consider a thin plate loaded in its plane. Assuming there are no stresses in the perpendicular direction, that means,

$$\sigma_x = 0, \quad \tau_{yz} = 0, \quad \tau_{zx} = 0 \qquad \text{[2.103]}$$

Consequently the elasticity matrix becomes

$$[\mathbf{D}] = \frac{E}{1-v^2} \begin{bmatrix} 1 & -v & 0 \\ -v & 1 & 0 \\ 0 & 0 & \dfrac{1-v}{2} \end{bmatrix} \qquad \text{[2.104]}$$

Plane strain Consider a plate of large thickness. Assuming the deformation to be in the plane of the plate and no displacements in the perpendicular direction, that means,

$$\varepsilon_x = 0, \quad \gamma_{yz} = 0, \quad \gamma_{zx} = 0 \qquad \text{[2.105]}$$

The corresponding elasticity matrix becomes

$$[\mathbf{D}] = \frac{E}{1-v} \begin{bmatrix} \dfrac{1-v}{1-2v} & \dfrac{v}{1-2v} & 0 \\ \dfrac{v}{1-2v} & \dfrac{1-v}{1-2v} & 0 \\ 0 & 0 & \dfrac{1}{2} \end{bmatrix} \qquad \text{[2.106]}$$

Thermal strains and stresses The derivations of strains and stresses until now have referred to mechanical loading. They can be extended, however, to include thermal loading as well. Thermal strains and stresses are induced by local heating or cooling in a body.

Thermal strains are defined by the equation

$$\{\varepsilon_{th}\} = \alpha \, \Delta T \begin{Bmatrix} 1 \\ 1 \\ 1 \\ 0 \\ 0 \\ 0 \end{Bmatrix} \qquad \text{[2.107]}$$

where ΔT denotes a local temperature difference and α is the thermal expansion coefficient. The resulting thermal stresses equal

$$\{\sigma\} = [D]\{\varepsilon_{th}\} \qquad\qquad [2.108]$$

Equilibrium

To explain the correlation between nodal displacements and forces in equilibrium, we use the principle of virtual work.

Consider the virtual work in the element shown in Figure 2.7(b). The virtual internal work equals

$$\delta W_i = \int_{(e)} \{\delta\varepsilon\}^T \{\sigma\} d(\text{vol}) \qquad\qquad [2.109]$$

The above can be modified by inserting the corresponding expressions of $\{\sigma\}$ and $\{\delta e\}$ as follows

$$\delta W_i = \int_{(e)} \{\delta u_e\}^T [B]^T [D][B]\{u_e\} d(\text{vol}) \qquad\qquad [2.110]$$

On the other hand, the virtual external work performed by nodal forces is

$$\delta W_F = \{\delta u_e\}^T \{F_e\} \qquad\qquad [2.111]$$

Based on the principle of virtual work, the internal and external works must be equal (see Section 2.1.4). Hence we obtain the equation of equilibrium in the element

$$\{F_e\} = \int_{(e)} [B]^T [D][B] d(\text{vol}) \cdot \{u_e\} \qquad\qquad [2.112]$$

The equation provides a correlation between nodal forces and nodal displacements of an element. The right hand integral represents the stiffness of the element.

Stiffness matrix   For computation purposes, the FE method includes the following $[K_e]$ matrix (so-called stiffness matrix) defined by

$$[K_e] = \int_{(e)} [B]^T [D][B] d(\text{vol}) \qquad\qquad [2.113]$$

whereby the equation of element equilibrium becomes

$$\{F_e\} = [K_e]\{u_e\} \qquad\qquad [2.114]$$

Global equilibrium   Until now we viewed the elements individually. Now we take a global view of all the elements combined in a FE model. The global equilibrium in this case is expressed by the equation

$$\{R_g\} = [K_g]\{u_g\} \qquad [2.115]$$

where $[K_g]$ is the global stiffness matrix and is the sum of element stiffness matrices

$$[K_g] = \sum [K_e] \qquad [2.116]$$

In Equation (2.115) the vector $\{u_g\}$ denotes the displacements of all nodal points within the entire body. Vector $\{R_g\}$ denotes the external forces, while internal forces $\{F_e\}$ disappear because their sum is zero. The global equilibrium condition is viewed by the FE method as a system of linear algebraic equations of nodal displacements $\{u_g\}$ unknown. For simple elastic problems with small displacements and the geometry being substantially unchanged, the global stiffness matrix is constant and Equation (2.115) is linear.

2.3.2 Nonlinear problems

We now turn to the FE solution of nonlinear problems of concern to fatigue design, namely, those that require more complex techniques:

(a) plastic deformation
(b) nonlinear geometry

Plastic deformation

Plastic deformation involves nonlinear stress-strain relations calling for nonlinear solution techniques. The solution method that follows is based on the incremental theory of plasticity.

The computation of displacements is based on incremental equilibrium. Consequently Equation (2.115) is replaced by the following equation

$$\{\Delta R_g\} = [K_{ep}]\{\Delta u_g\} \qquad [2.117]$$

$[K_{ep}]$ is a global elastic-plastic stiffness matrix which comprises nonlinear element stiffness matrices

$$[K_{ep}] = \int_{(e)} [B]^T [D_{ep}][B] d(\text{vol}) \qquad [2.118]$$

The elasticity-plasticity matrix $[D_{ep}]$ within the integral is derived below.

Solid Mechanics

A differentiation of the yield function, Equation (2.61), of plasticity theory, produces the expression

$$dF = \left\{ \frac{\partial F}{\partial \boldsymbol{\sigma}} \right\}^{T} d\sigma + \frac{\partial F}{\partial \kappa} d\kappa = 0$$

[2.119]

On the other hand, the incremental strain component of elastic-plastic deformation equals

$$\{\Delta \boldsymbol{\varepsilon}\} = \{\Delta \boldsymbol{\varepsilon}^{e}\} + \{\Delta \boldsymbol{\varepsilon}^{p}\} = \{\Delta \boldsymbol{\sigma}\}[D_{e}]^{-1} + \Delta \lambda \left\{ \frac{\partial f}{\partial \boldsymbol{\sigma}} \right\}$$

[2.120]

where $[D_{e}]$ is the elasticity matrix defined by Equation (2.102). Combining both equations equals

$$\left\{ \begin{array}{c} \Delta \boldsymbol{\varepsilon} \\ 0 \end{array} \right\} = \left[\begin{array}{cc} [D_{e}]^{-1} & \left\{ \dfrac{\partial f}{\partial \boldsymbol{\sigma}} \right\} \\ \left\{ \dfrac{\partial f}{\partial \boldsymbol{\sigma}} \right\} & -H^{*} \end{array} \right] \left\{ \begin{array}{c} \Delta \boldsymbol{\sigma} \\ \Delta \lambda \end{array} \right\}$$

[2.121]

where parameter H^{*} is defined as

$$H^{*} = -\frac{\partial f}{\partial \kappa} d\kappa \frac{1}{d\lambda}$$

[2.122]

The stress-strain correlation can be expressed by the following equation

$$\{\Delta \boldsymbol{\sigma}\} = [D_{ep}]\{\Delta \boldsymbol{\varepsilon}\}$$

[2.123]

Equation (2.121), in conjunction with the Equation (2.123), produces the expression of elasticity-plasticity matrix

$$[D_{ep}] = [D_{e}] - [D_{e}] \frac{\left\{ \dfrac{\partial f}{\partial \boldsymbol{\sigma}} \right\} \left\{ \dfrac{\partial f}{\partial \boldsymbol{\sigma}} \right\}^{T} [D_{e}]}{H^{*} + \left\{ \dfrac{\partial f}{\partial \boldsymbol{\sigma}} \right\}^{T} [D_{e}] \left\{ \dfrac{\partial f}{\partial \boldsymbol{\sigma}} \right\}}$$

[2.124]

The derivation of parameter H^{*} follows from the analysis of one-dimensional specimen in axial loading. The applicable yield function equals

$$F = \sigma_{1} - \sigma_{e}(\kappa) = 0$$

[2.125]

Yield parameter κ is defined as the plastic work

$$d\kappa = \sigma_{e} d\varepsilon_{p}$$

[2.126]

Differentiating yield function F we obtain

$$-\frac{\partial F}{\partial \kappa} = \frac{1}{\sigma_e} \cdot \frac{d\sigma_e}{d\varepsilon_p} = \frac{H'}{\sigma_e} \qquad [2.127]$$

where H' is the slope of one-dimensional stress-strain curve. It follows that

$$H^* = H' \qquad [2.128]$$

The elasticity-plasticity matrix derived above, Equation (2.124), is used in elastic-plastic analysis after the input of yield function. One can choose either one based on von Mises or on Tresca theory. For example, von Mises yield function produces the following elasticity-plasticity matrix[10]

$$[D_{ep}] = [D_e] - \frac{\dfrac{3}{2\sigma_e^2}\dfrac{E}{1+v}}{1+\dfrac{2}{3}(1+v)\dfrac{H'}{E}} \begin{bmatrix} s_x^2 & & & & & \text{sym.} \\ s_x s_y & s_y^2 & & & & \\ s_x s_z & s_y s_z & s_z^2 & & & \\ s_x \tau_{xy} & s_y \tau_{xy} & s_z \tau_{xy} & \tau_{xy}^2 & & \\ s_x \tau_{yz} & s_y \tau_{yz} & s_x \tau_{yz} & \tau_{xy}\tau_{yz} & \tau_{yz}^2 & \\ s_x \tau_{zx} & s_y \tau_{zx} & s_z \tau_{zx} & \tau_{xy}\tau_{zx} & \tau_{yz}\tau_{zx} & \tau_{zx}^2 \end{bmatrix} \qquad [2.129]$$

Nonlinear geometry

Nonlinear geometry pertains to problems which include geometrical changes caused by loading. We consider two cases: (a) large displacements affecting the geometry, and (b) contact problems of multiple bodies. The solutions proceed in steps, applying incremental loading as highlighted below.

Large displacements   The nonlinearity results from changes in matrices [B] of the finite element model. The solution, based on strain increments, is defined as

$$\{\Delta\boldsymbol{\varepsilon}\} = [\mathbf{B}^*]\{\Delta\mathbf{u}_e\} \qquad [2.130]$$

Matrices [B*] are usually divided into two parts, linear and nonlinear,

$$[\mathbf{B}^*] = [\mathbf{B}_l] + [\mathbf{B}_n] \qquad [2.131]$$

resulting in a nonlinear global stiffness matrix [K*]. (See References 8 and 9.)

Contact problems of multiple bodies   The nonlinearity results from the changing contact area between the bodies under loading. The solution is reached through a simulation of the contact interface using gap elements to connect nodes in contact. Since the gap element stiffness varies (in compression it has a given value while in tension it equals zero), the contact area under loading is represented by a variable stiffness matrix [K*]. The finite element solution proceeds in iterative steps with a variable global stiffness matrix, activating and deactivating the respective gap elements.

Newton's method of iterations

The process of solving nonlinear problems ended with setting up nonlinear stiffness matrices ready for computation. In accordance with the FE method for the solution of an equilibrium equation, incorporating the nonlinear stiffness matrix, it is necessary to obtain the nodal displacements. This is followed by a computation of strains and stresses.

The computation is generally done by the Newton method of iterations. The equilibrium equation is presented in the form

$$\left\{\Delta R_g\right\} = \left[K_{g,n}\right]\left\{\Delta u_g\right\} \qquad [2.132]$$

where global matrix $[K_{g,n}]$ is nonlinear. At a given i-iteration, displacements $\{\Delta u_g\}^i$ are computed by the recurrence equation

$$\left\{\Delta R_g\right\} = \left[K_{g,n}\right]^{i-1}\left(\left\{\Delta u_g\right\}^i - \left\{\Delta u_g\right\}^{i-1}\right) + \left\{\Delta F_g\right\}^{i-1} \qquad [2.133]$$

The recurrent computation ceases after a desired accuracy is reached.

The iteration process, solving a one-dimensional problem, is presented graphically in Figure 2.8. Figure 2.8(a) shows the Newton-Raphson method while Figure 2.8(b) pre-

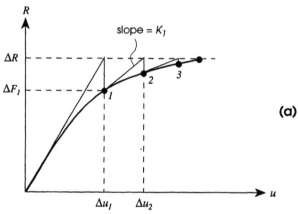

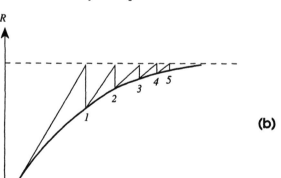

Figure 2.8 Newton iteration methods; (a) Newton-Raphson Method; (b) modified Newton Method

sents the modified Newton method with constant stiffness matrix. As seen in the figure, vector $\{\Delta F_g\}^{i-1}$ is a recurring term approaching $\{\Delta R\}$.

REFERENCES

1. Biezeno, C.B., and R. Grammel. 1954. *Engineering Dynamics*. London: Blackie and Son.
2. Lanczos, C. 1964. *The Variational Principles of Mechanics*. Toronto: Toronto University Press.
3. Krempl, E. 1974. "The Influence of State of Stress on Low-Cycle Fatigue of Structural Materials." ASTM STP 549.
4. *ASME Boiler and Pressure Vessel Code*. 1989. Section 8, Rules for Construction of Pressure Vessels. New York: ASME.
5. Hill, R. 1971. *The Mathematical Theory of Plasticity*. London: Oxford University Press.
6. Mendelson, A. 1968. *Plasticity: Theory and Application*. New York: Macmillan.
7. Martin, J. 1975. *Plasticity: Fundamentals and General Results*. 3rd ed. Cambridge, Massachusetts: MIT Press.
8. Zienkiewicz, O.C. 1977. *The Finite Element Method*, 3rd ed. London: McGraw-Hill.
9. Bathe, K.J. 1982. *Finite Element Procedures in Engineering Analysis*. Englewood Cliffs, NJ: Prentice-Hall.
10. Desai, C.S., and J.F. Abel. 1972. *Introduction to the Finite Element Method*. New York: Van Nostrand Reinhold.

3

STRESS METHOD

The tendency to diagnose machine part failures, since the beginning of machine design analysis, has been on examination of the stresses. The main concern then was, as it is now, the stress magnitude, but with one difference—no consideration was given to the crucial effects of fluctuating loads. In the past, to guard machine parts against failure, only standard static properties obtained from the tests were used: either the yield point or the ultimate strength of the metal. These analyses, based on static forces, neither predicted nor explained with any accuracy the failures occurring at fluctuating loadings.

With time, objective observations proved that a machine part can withstand static stresses of high magnitude but will fail if these stresses fluctuate. It was confirmed that the safe limits in design analysis of a machine part subjected to fluctuating loading must be below those applicable to static loading. This discovery prompted the definition of a new property called the fatigue limit, adding it to those solely used for analysis before— the yield point and the ultimate strength of the metal. Today the fatigue limit (the history of which goes as far back as the nineteenth century) has become a necessary tool.

The main consideration in designing a part which is subject to fluctuating loading is to determine the extent of the desired number of cycles to fatigue failure; that is, should it have an infinite fatigue life span or a finite number of cycles? Either can be considered with one fact in mind. For an infinite fatigue life stresses must be kept below the fatigue limit, because stresses above the fatigue limit will cause a failure after a given number of cycles. Different constraints inherent in the design determine the magnitude of stresses and these are the factors that dictate the proper choice. The Stress Method utilizes the fact that there exists a correlation between the working stresses and the number of cycles within the finite life. It is based on the assumption that a part behaves elastically ignoring the occurrence of plastic deformation.

3.1
TESTS AND TEST RESULTS

Since the beginning of machine design, the experimental data has been the practical basis for forecasting failures of machine parts. Most of the methods for forecasting are based on experimental results and the Stress Method discussed in this chapter is no exception. The results of fatigue tests provide a tool for prediction of life expectancy: the correlation of number of fluctuations to the magnitude of applied stresses.

One should note that the rate of load fluctuations during tests is a factor that is virtually ignored. Experience shows that within the range of about 1,000 to 10,000 cycles per minute, no perceptible effect of it on the correlation between number of load cycles and stresses has been recorded.

There are standard fatigue tests which include rotating-beam and axial-loading tests (see below) to provide the limit values of normal stresses. The results obtained can also be used to derive other limit values as, for instance, the torsional shear fatigue limits. To interpret the test results, however, one must properly use the available statistical tools.

3.1.1 Standard fatigue tests

Design of machine parts uses the fatigue information derived from standard tests where stresses fluctuate at constant amplitudes—due to practical constraints—even though most machine parts are subject to fluctuating stresses with random amplitude and mean. See, for instance, the loading history shown in Figure 3.1. This simplification can be remedied subsequently using available analytical tools, as will be explained later.

The history of standard fatigue tests goes back to Wohler[1] who designed and built the first rotating-beam test machine that produced fluctuating stresses of a constant amplitude in test specimens. See Figure 3.2. The specimens, two railroad axles mounted

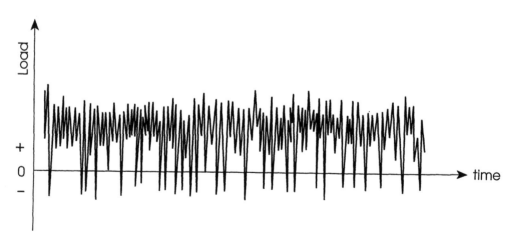

Figure 3.1 Recording of fluctuating load in a tractor axle.

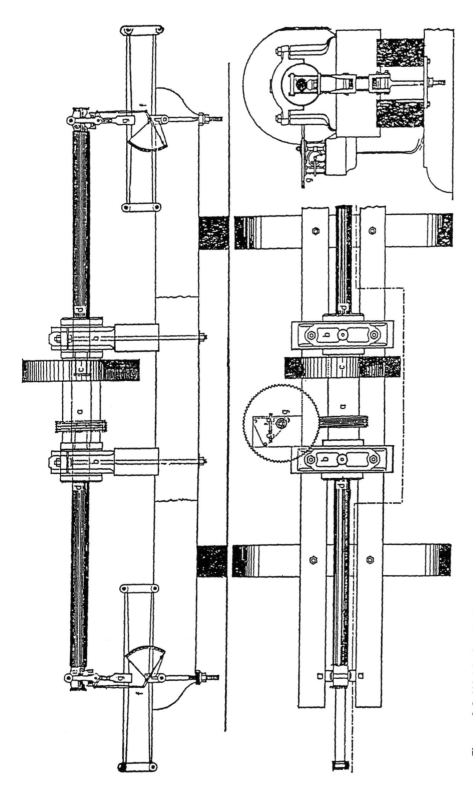

Figure 3.2 Wohler's rotating-beam fatigue testing machine. (Wohler, A., see Reference 1.)

symmetrically and bent by constant spring forces, were rotated at a constant speed until failure. In these tests Wohler established a material property, known today as the "fatigue limit." He discovered that a steel specimen is safe as long as the stresses are kept below a certain limit. The rotating-beam test machines used today in modern technology are built on the same principle as that of Wohler's, subjecting the specimen to fluctuating bending stresses. There are other test machines, such as the axial-loading machines, which subject the specimen to fluctuating axial stresses. In the following discussion, we consider the two modes of testing—rotating beam and axial loading.

Rotating-beam testing

The test procedure   The machine most commonly used is the constant-moment fatigue testing machine. See Figure 3.3. Motor M is flexibly coupled to a rotating test specimen S which is subjected to a constant bending moment over its entire length. All bearings are self-aligning to ensure that only radial loads are applied to the specimen. Counter R registers the number of revolutions. The electric contactor stops the motor when a specimen breaks.

This machine works in accordance with the rotating bending principle that, as the specimen rotates around its bent axis, it is subjected to cyclic stresses of a constant amplitude. The specimen's extreme fibers experience a stress which alternates with time between maximum and minimum. At each half revolution it changes from tension to an equal compression. The cycle is repeated at the frequency of rotation. The stress follows a sinusoidal law. See Figure 3.4. A substantial number of specimens are tested at different stress amplitudes to register the failure points.

The specimen form shown in Figure 3.5 is the most commonly used in standard tests of this nature. It is round with a shallow fillet of 250 mm radius and a minimum diameter of 8 mm. The large fillet radius renders the specimen free of stress concentrations and a heat treatment frees it of residual stresses. The highest stresses, where the breakage is predicted to occur first, are present at the midsection.

Presentation of results   The methods of graphically correlating the test data have been perfected with time. Let us follow from Figure 3.6 the evolution of presenting cyclic stress amplitude S versus number of cycles to failure N, for ferrous metals. Original plots of S–N line were in a linear scale (see Figure 3.6.a) evolving later into diagrams in a semi-log scale (Figure 3.6.b) and then log-log scale (Figure 3.6.c). All three

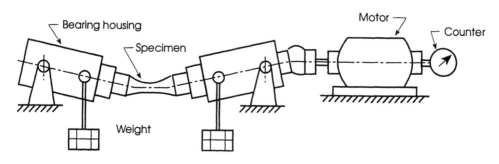

Figure 3.3 Moore rotating-beam fatigue testing machine.

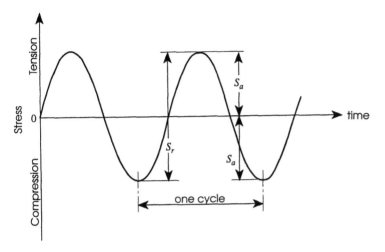

Figure 3.4 Stress pattern in a specimen during a rotating-beam test.

plots reflect the character of $S–N$ line: it starts as a sloping curve changing into a horizontal asymptote with N becoming infinite. The transformation is best identified in the log-log diagram (Figure 3.6.c) where an abrupt knee, marking the beginning of asymptote, can be easily observed. For steels the knee generally occurs at N less than 5×10^6 cycles. For nonferrous metals and alloys and some high strength alloy steels the $S–N$ curve behaves differently. Because of the length of the testing period required to reach the fatigue limit, the $S–N$ curve is commonly plotted to extend up to 5×10^8 cycles, despite knowing that the curve may not slope into a horizontal line at all.

The ASTM standard nomenclature[2] refers to a value of stress amplitude S (located in the sloping part) as Fatigue Strength and the stress level of the asymptote as the Fatigue Limit (defined below). Values of fatigue strength and fatigue limit found in the literature are as a rule statistical mean values of test results. In some instances a more detailed statistical evaluation may be applicable, as will be explained in Section 3.1.3.

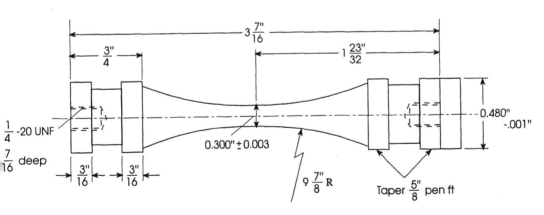

Figure 3.5 Rotating-beam test specimen.

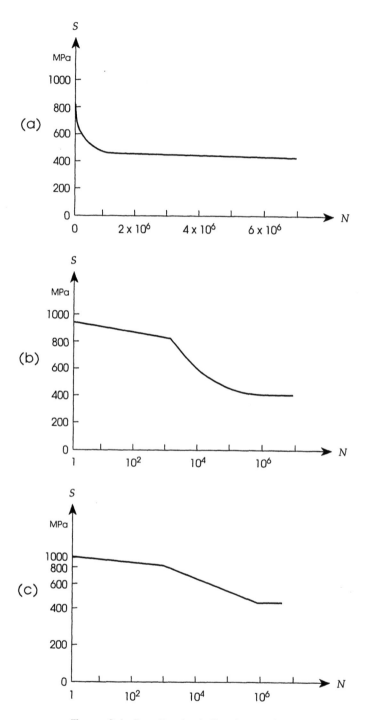

Figure 3.6 Results of rotating-beam fatigue tests
for steel: (a) S-N curve in linear coordinates;
(b) S-N curve in semi-log coordinates;
(c) S-N curve in log-log coordinates.

Axial-loading testing

Axial loading can be produced by hydraulic, electromagnetic, and mechanical forces. The following is description of the test procedure performed on a machine operated by hydraulic forces.

The test procedure Figure 3.7 shows a modern test machine operating by hydraulic forces and controlled by electrical signals. The machine loads the specimen (a) by means of hydraulic actuator (b). Changing the direction of the load while the machine is in operation is done by reversing the oil flow by electrical command. That is, the oil flow to and from the actuator is controlled by servo-valve (c) which operates by means of an electomagnetic solenoid. Figure 3.8 shows a standard specimen for axial loading free of stress concentrations and residual stresses.

The purpose of the machine is to subject the specimen simultaneously to two types of axial stresses caused by the respective loads: cyclic and steady. Figure 3.9 is a diagram which shows a superposition of both cyclic and steady stresses. The axial-loading machine has certain disadvantages—large forces to achieve the necessary stresses and difficulty in lining up the specimen with the load. An eccentric load, due to improper mounting, may bend the specimen and cause a nonuniform stress distribution which, in turn, can lead to erroneous test data.

Presentation of results To explain the test results we turn to the following mathematical expressions. Let us denote maximum stress S_{max}, minimum stress S_{min}, mean

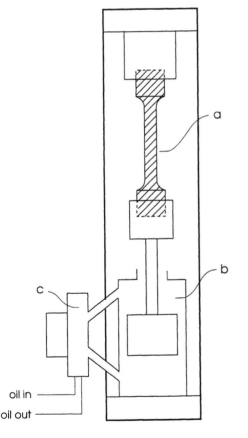

Figure 3.7 Electro-hydraulic axial-load fatigue testing machine: (a) specimen; (b) hydraulic actuator; (c) servo-valve

oil in

oil out

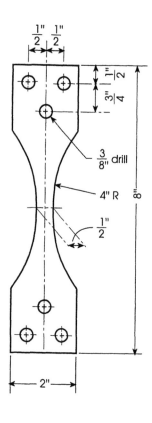

Figure 3.8 Axial-load
test specimen.

stress S_m and cyclic stress amplitude S_a. The above mentioned steady stress, caused by steady load, equals the mean stress

$$S_m = \frac{S_{max} + S_{min}}{2} \qquad [3.1]$$

The amplitude S_a equals

$$S_a = \frac{S_{max} - S_{min}}{2} \qquad [3.2]$$

Now we can present the ratios needed for test evaluation and their relation: the ratio of the cyclic stress amplitude to the mean stress

$$A = \frac{S_a}{S_m} \qquad [3.3]$$

and the ratio of the minimum stress to the maximum stress

$$R = \frac{S_{min}}{S_{max}} \qquad [3.4]$$

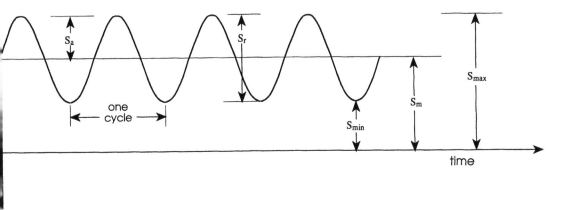

Figure 3.9 Stress pattern in a specimen during a constant amplitude axial-load test.

The relation between A and R then becomes

$$A = \frac{1-R}{1+R}, \qquad R = \frac{1-A}{1+A} \qquad\qquad [3.5]$$

Thus, designing the tests with mean stress S_m changing systematically, the effect of the steady stress on the fatigue life is revealed. Figure 3.10 shows typical axial-loading test results for various stress ratios R.[3] Note that the $S-N$ curve for $R = -1$ corresponds to cyclic stresses with zero mean.

The magnitude of fatigue limits in axial-loading tests is below that of rotating-beam tests, as shown in Figure 3.11 where results of the two tests are compared. A high statistical probability of existence of internal flaws in materials gives rise to unpredictable stress concentrations. In testing, this fact affects differently the specimens tested in axial-loading and rotating-beam machines: the occurrence of a failure can be expected sooner in the axial-loading test than in the rotating-beam test. An explanation can be found in the fact that in axial-loading tests the entire cross-sectional area of the specimen is stressed, while in rotating-beam tests only the area in the vicinity of surface is stressed. See Figure 3.12.

Fatigue limit

The fatigue limit (also called endurance limit) is defined as the maximum value of stress amplitude at zero mean stress which can be repeated an infinite number of times on a test specimen without causing a failure. This value forms a mechanical property specific for each material. The fatigue limit depends on the loading mode (the result is different for specimens subjected to rotating bending, axial loading, or torsion) and it is influenced by the size and surface condition of the specimen. The test results used in design of machine parts combine these effects.

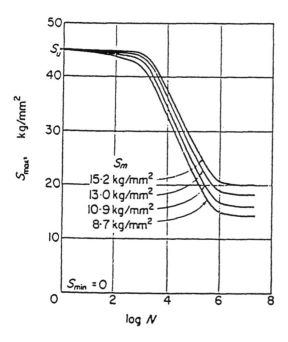

Figure 3.10 Results of axial-load fatigue tests with Alclad specimens for various mean stresses S_m. (Weibull, W., see Reference 3. Copyright ARD NATO.)

To determine the fatigue limit of a typical machine part, we need to utilize the data of fatigue limits of standard specimens in rotating-beam tests. Our typical machine part has an arbitrary size and surface condition and an arbitrary loading mode (bending, axial, or torsion). The standard specimen has a polished surface and 8 mm diameter at its neck. The fatigue limit of the machine part, S_f, is expressed by the following empirical correlation

$$S_f = k_a k_b k_c S_f'$$ [3.6]

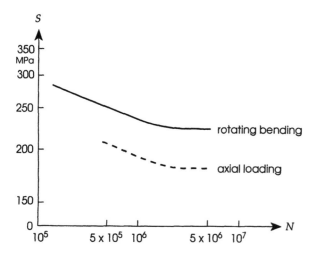

Figure 3.11 Results of two fatigue test modes with steel specimens.

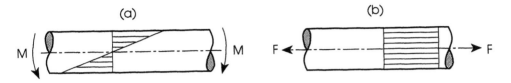

Figure 3.12 Stress distribution in two fatigue test modes:
(a) rotating-beam test; (b) axial-loading test.

where S'_f is the fatigue limit of the standard specimen and k_a, k_b and k_c are modification factors defined as follows: k_a—loading mode factor; k_b—size effect factor; and k_c—surface roughness factor.

Loading mode factor Let us take three standard specimens and subject each one to a different kind of fatigue testing, that is, to rotating beam, to axial loading, and to torsion. As mentioned above, although the three specimens are the same, the fatigue limits will differ for each. Let us designate the fatigue limit of the specimen subjected to rotating-beam testing as a measuring constant as related to the other two fatigue limits. Their ratio to this constant is known as the loading mode factor k_a.

The loading mode factor k_a for axial loading as provided by tests varies from 0.8 to 1.0 for all kinds of metals. It is an accepted practice in machine design to use a mean value of 0.9. [Mischke[4] shows the value for steels with $S_u < 1500$ MPa (220 ksi) is 0.923, and for steels with $S_u > 1500$ MPa equals 1.0.]

Let us consider the loading mode factor for torsional loading. For computational convenience the standard specimen used here is in the form of a round bar. The shear stress distribution has a linear shape and the highest stress occurs on the bar's surface. See Figure 3.13. Assume a limit condition where the highest stress equals the fatigue limit τ_f, as in the figure. Let us introduce the equivalent stress S_{eq}, based on von Mises criterion. Derived from Equation (2.49) it equals

$$S_{eq} = \sqrt{0 + 3\tau^2} \qquad [3.7]$$

One can assume that S_{eq} is equal to S'_f, the bending fatigue limit of the same round bar. Now the latter equation becomes

$$S'_f = \sqrt{0 + 3\tau_f^2} \qquad [3.7']$$

From here we can compute the fatigue limit in torsion

$$\tau_f = \frac{1}{\sqrt{3}} S'_f = 0.577 S'_f \qquad [3.8]$$

Equation (3.8) is confirmed by tests. The derived value of 0.577 is the loading mode factor for torsional loading. A similar derivation based on the maximum shear stress theory produces a value of 0.5.

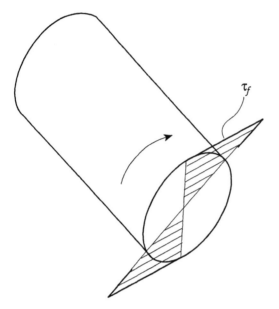

τ_f

Figure 3.13 Shear stress distribution in a round
bar subject to torsion.

Figure 3.14 compares data from tests with plots based on von Mises and Tresca
theories.[5] Both theories seem valid, although some disparity is noted. The derivation
based on von Mises criterion is usually preferred.

To summarize the three loading mode factors:

<table>
<tr><td>for rotating bending</td><td>$k_a = 1.0$</td><td>[3.9]</td></tr>
<tr><td>for axial loading</td><td>$k_a = 0.9$</td><td></td></tr>
<tr><td>for torsional loading</td><td>$k_a = 0.577$</td><td></td></tr>
</table>

Size effect factor   As noted above, the size of specimens must be considered in
defining the fatigue limit. Experiments show that for rotating beam and for torsion the
value of fatigue limit changes inversely to the diameter of a specimen, while in axial
loading the size has no effect.

For designing purposes in rotating beam and in torsion the following empirical
factors are in use

<table>
<tr><td>for $d < 8$mm</td><td>$k_b = 1.0$,</td><td></td></tr>
<tr><td>for $8 < d < 50$mm</td><td>$k_b = 0.85$</td><td>[3.10]</td></tr>
</table>

where d is the diameter of the specimen. A recent study by Mischke (see Reference 4)
produced slightly different factors

$$k_b = \left(\frac{d}{8.0}\right)^{-0.1133} \quad \text{for } 2.5 < d < 50\text{mm} \tag{3.11}$$

$$k_b = 0.85 \qquad \text{for } d > 50\text{mm}$$

where d is measured in mm.

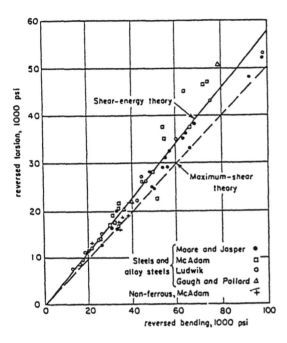

Figure 3.14 Relation between fatigue limits in bending and in torsion. ("ASME Handbook, Metals Engineering Design," see Reference 5. Reprinted by permission.)

The size effect factor for axial loading, independent of size d, equals

$$k_b = 1.0 \qquad\qquad\qquad\qquad\qquad\qquad [3.12]$$

Note that the distribution of stresses in axial loading is constant, as shown in Figure 3.12(b) while in rotating beam or in torsion there is a stress gradient, as shown in Figures 3.12(a) and 3.13. In the latter cases (rotating beam and torsion) there is an inverse relationship between the size and the stress gradient—an increased diameter lowers the gradient. This may explain the fact that k_b for axial loading equals 1, while for for rotating beam and torsion it is less than 1 for larger diameters.

Surface roughness factor As above, it can be demonstrated that the surface roughness affects the fatigue limit. The stresses in the specimen are computed on the basis of the general geometry. The surface roughness or the local irregularities are the high stress concentration points where the fatigue failures generally originate. The fatigue limit for a specimen with a polished surface, free from such defects, has therefore a higher value. There is an inverse relationship between the fatigue limit and the magnitude of the irregularity—the limit decreases with greater roughness. One must note that this applies only to fluctuating loading. In static tests neither the yield point nor the ultimate strength are affected by the surface roughness.

Figure 3.15 presents the surface roughness factor as a function of the surface finishing process such as machining, grinding, rolling and forging.[6-8] Factor k_c is different in each case because each tooling produces a different surface roughness. As shown, it varies from 0.3 to about 0.95.

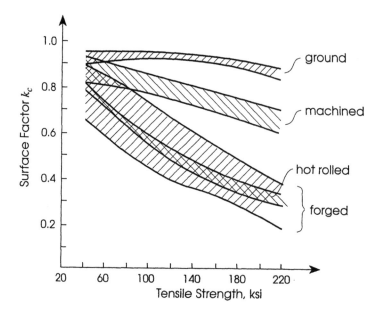

Figure 3.15 Surface factors versus tensile strength for steel. (Data from Faires, V. M., see Reference 6; Heywood, R. B., see Reference 7; Juvinal, R. C., see Reference 8.)

Ratio of fatigue limit to tensile strength   This ratio is another empirical factor that is useful in fatigue analysis. A close examination of test data of steel specimens shows a direct correlation between the rotating beam fatigue limit (obtained from standard tests) and the ultimate tensile strength of the metal. Figure 3.16 presents this correlation in a graphical form (see Reference 7). For steels having a tensile strength up to 1400 MPa (200 ksi), one finds the ratio to be about 0.5. For steels above 1400 MPa the ratio falls below 0.5. See Chapter 6.

For design purposes, one can use the following empirical formulas

$$S'_f = 0.5\ S_u \text{ for steels with } S_u < 1400\text{MPa (200ksi)},$$ [3.13]

$$S'_f = 100 \text{ ksi for steels with } S_u > 1400\text{MPa (200ksi)}$$

Consequently, an empirical formula for the fatigue limit of steel with $S_u < 200$ ksi can be derived as

$$S_f = k_a k_b k_c \frac{S_u}{2}$$ [3.14]

Stress concentration caused by notches and holes

Consider test specimens with body geometry that includes notches or holes. Under loading, the stresses concentrate and rise with a sharp gradient to a peak at the notch

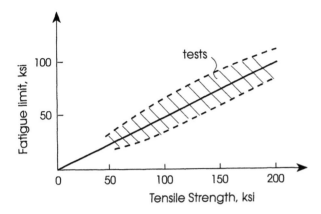

Figure 3.16 Rotating-beam fatigue limit versus tensile strength for steel. (Data from Heywood, R. B., see Reference 7).

border. See Figure 3.17. This phenomenon is called stress concentration and is a function of the curvature of the notch—as the radius of curvature gets smaller, the stress concentration increases. The locus of stress concentration represents the danger point. Comparing test results of a notched specimen and a smooth one, we witness a contradiction. After both specimens were subjected to the same number of cycles, the failure occurred in the notched one at a lower load level because of the sharp stress gradient caused by the uneven stress distribution.

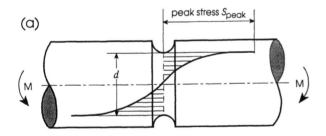

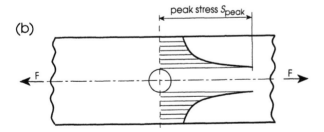

Figure 3.17 Stress concentration:
(a) in a bending specimen with a circular notch;
(b) in an axial-load specimen with a hole.

Peak stresses at a notch are first computed disregarding the notch curvature, arriving at nominal stresses and then correcting the values by means of a stress concentration factor,

$$\sigma_{\text{peak}} = K_t \sigma_{\text{nom}} \qquad [3.15]$$

Stress concentration factor K_t above is based on the elasticity theory. A discrepancy found between the theoretical and experimental data demands a substitution of factor K_t by a more accurate measurement observed in fatigue tests with a notched specimen, called the fatigue notch factor K_f.

To derive the fatigue notch factor from tests, let us consider a test specimen equipped with a notch similar to the part shown in Figure 3.17(a). The specimen is subjected to a rotating beam test. Its theoretical stress concentration factor is defined by the equation

$$K_t = \frac{\text{theoretical peak stress in notched specimen}}{\text{nominal stress in notched specimen}} \qquad [3.16]$$

where the nominal stress ignores the stress concentration. The fatigue notch factor is defined by the equation

$$K_f = \frac{\text{true peak stress in notched specimen}}{\text{nominal stress in notched specimen}} \qquad [3.17]$$

When the notched specimen fails in the fatigue test, its peak stress corresponds exactly to the stress in a smooth specimen. Therefore Equation (3.17) can be restated

$$K_f = \frac{\text{fatigue limit of smooth specimen}}{\text{fatigue limit of notched specimen}} \qquad [3.18]$$

This Equation (3.18) is used to obtain values of K_f from fatigue tests.

Experience shows that the fatigue notch factor for metals is less or equal to the theoretical stress concentration factor,

$$K_f \leq K_t \qquad [3.19]$$

It appears that inequality (3.19) is a result of localized plastic deformation, taking place within the notch. An empirical correlation based on the notch curvature and material properties has the form (see Reference 7)

$$K_f = 1 + \frac{K_t - 1}{1 + \sqrt{\dfrac{a}{r}}} \qquad [3.20]$$

where r is notch radius and a is a material property constant having the dimension of length. Experimental values of a are presented in Figure 3.18.[9] Equation (3.20) is known as Neuber's formula. We can simplify it by stating

$$q = \frac{1}{1 + \sqrt{\dfrac{a}{r}}}$$

[3.21]

whereby we obtain

$$K_f = 1 + q(K_t - 1)$$

[3.22]

and q is called the notch sensitivity. Figure 3.19 shows computed values of q for steel and aluminum.

A different approach can be taken to determine the correlation of the fatigue notch factor K_f and the theoretical stress concentration factor K_t, starting with a definintion of the ratio

$$q = \frac{\text{effective increase of stress due to notch}}{\text{theoretical increase of stress due to notch}}$$

[3.23]

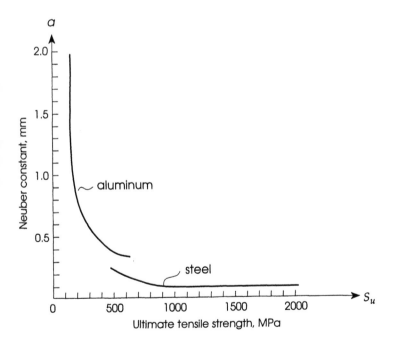

Figure 3.18 Neuber constants for steel and aluminum. (Data from Grover, H. J., see Reference 9.)

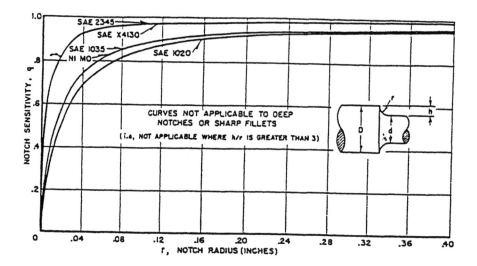

Figure 3.19 Notch sensitivity curves. ("ASME Handbook, Metals Engineering Design," see Reference 5. Reprinted by permission.)

which may be written[10] as

$$q = \frac{K_f - 1}{K_t - 1}$$
[3.24]

Equation (3.24) leads to the same results as before, see Equation (3.22).

3.1.2 Statistical analysis of fatigue data

To obtain meaningful engineering data, that is, values that can be reliably applied to machine design, a large number of specimens must be tested. The tests are followed by a statistical treatment of data collected. Even though the same standard specimens are used in these tests, the results show a wide dispersion of data. It is important to choose a suitable procedure to provide maximum information with a minimum number of tests. The reason for data dispersion lies in the fact that even the same standard specimens have different geometrical micro irregularities of surfaces. As previously mentioned, local stress concentrations cause different fatigue lives for each specimen at the same load level.

For illustration, consider a typical test procedure at a constant stress level. Let us denote

$$x = \log N$$
[3.25]

Plotting fatigue failures as a function of N, we obtain a histogram. See Figure 3.20(a). For statistical interpretation it is necessary to substitute the histogram with a continuous distribution curve. For this we have a number of statistical expressions available, for example, Gauss distribution function, Weibull distribution function, and others.[11] The distribution function of Gauss (also called the normal distribution function) is expressed by the equation

$$f(x) = \frac{1}{\sqrt{2\pi}\,\sigma}\; e^{-\frac{1}{2}\left(\frac{x-\bar{x}}{\sigma}\right)^{-2}}$$

[3.26]

where $\bar{x}$ is a mean value of log N and σ denotes standard deviation. The probability of

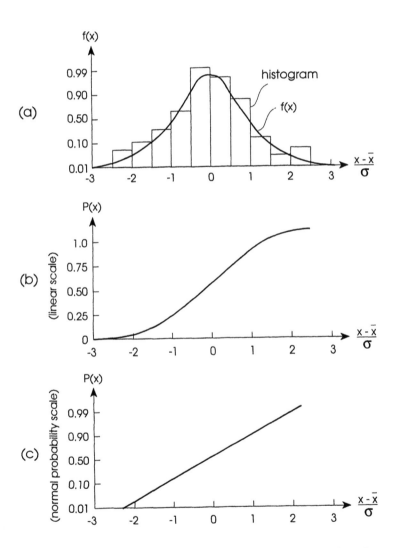

Figure 3.20 Plots of the normal distribution function and the cumulative frequency: (a) histogram and a matching distribution function f(x); (b) cumulative frequency function P(x); (c) function P(x) in a normal probability scale.

failure at N (so-called the cumulative frequency from 0 to x) is defined by the equation

$$P = \int_0^x f(x)dx = \int_0^x \frac{1}{\sqrt{2\pi}\,\sigma}\, e^{-\frac{1}{2}\left(\frac{x-\bar{x}}{\sigma}\right)^2} dx \qquad [3.27]$$

From Figure 3.20 we see the plotting of normal distribution $f(x)$ and cumulative frequency $P(x)$. We can transform the P scale in the P–x plot so that P versus x becomes a straight line. See Figure 3.20(c). This has a practical application in the treatment of fatigue data, as shown below.

The Weibull distribution function is expressed by equation

$$f(x) = b\frac{(x - x_0)^{b-1}}{a^b}\, e^{-\left(\frac{x-x_0}{a}\right)^b} \qquad [3.28]$$

where x_0 is an expected minimum value of x, a is a scale factor, and b is a slope parameter to be explained later. The corresponding cumulative frequency equals

$$P = \int_0^x b\frac{(x - x_0)^{b-1}}{a^b}\, e^{-\left(\frac{x-x_0}{a}\right)^b} dx = 1 - e^{-\left(\frac{x-x_0}{a}\right)^b} \qquad [3.29]$$

In general, the scatter in fatigue lives is modeled better by the Weibull distribution than by the Gaussian distribution.

The P–S–N *correlation*

The standard S–N curve (Figure 3.6) is based on mean values, that is, it corresponds to a 50 percent probability of failure ($P = 0.5$). Having discussed the necessity of statistical analysis of a wide data dispersion, the need for other values of P is now apparent. In other words, it is necessary to add P to the S–N correlation, providing an expression for P–S–N dependence. An example is shown in Figure 3.21. This is a typical P–S–N diagram comprised of lines corresponding to different constant values of P in S–N coordinates and forming confidence bands.

To illustrate the P–S–N correlation, let us consider the results obtained from tests at four stress levels. See Figure 3.22. Plotting fatigue failures taken from the data, we first obtain a number of histograms which are then substituted with Gaussian distribution curves. See Equation (3.26). These curves allow us to define probability P, as per Equation (3.27), at each stress level S. In graphical form (connecting the same value points of P at different stress levels) we obtain S–N curves with constant values of P as parameters. Thus stress S is expressed as function $S(N,P)$, providing a P–S–N correlation. Now the function may be stated in a reciprocal form $P(N,S)$, that is, probability of failure P versus fatigue life N, using S as parameters. Its graphical representation is obtained by plotting P–N curves at constant values of S on a log-normal probability paper, where abscissa N is in logarithmic scale and ordinate P is in normal-probability

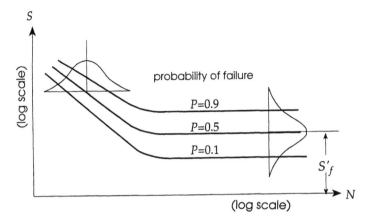

Figure 3.21 Probability distribution of *S-N* data.

scale. Figure 3.23 presents the P–S–N diagram where failure data are grouped according to stress S. (See Reference 3.)

To emphasize the effects of using different distribution functions, let us substitute the test histograms with Weibull distribution curves, Equation (3.28). Figure 3.24, where ordinate P is in Weibull probability scale, is a diagram showing that failure points at constant S fit closer straight lines. The value of slope of the straight line defines parameter b in the equation.

Determination of fatigue limit

As has been already noted, a fatigue strength for a given fatigue life N is defined as the value of stress amplitude S, located in the sloping part of S–N curve, and fatigue limit is defined as the limiting value of the stress amplitude when N becomes very large. To determine a fatigue limit experimentally, the test results are evaluated statistically using either the data of specimens that survived (run-outs) or of those that failed. The

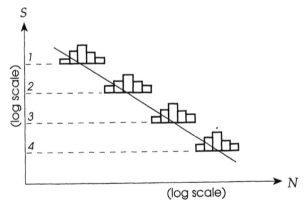

Figure 3.22 Fatigue life distribution at four stress levels.

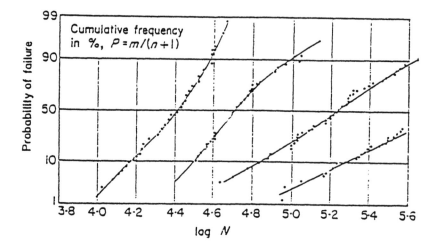

Figure 3.23 Log-normal probability plot *P* versus *N* with stress *S* as parameter. (Weibull, W., see Reference 3. Copyright ARD NATO.)

stress levels used on test specimens are based on preliminary knowledge of their estimated fatigue strength at desired life expectancy.

One method of testing is known as the "staircase method" or the "up-and-down method." The staircase method is structured so that the result of the previous test forms the basis for the subsequent test and so on until the end of the series. Its application is twofold, for the determination of fatigue strength distribution as well as of fatigue limit. Let us consider this test procedure as applied to the fatigue limit. A specimen is tested starting at a predetermined stress level. That level is lowered at the next try if the spec-

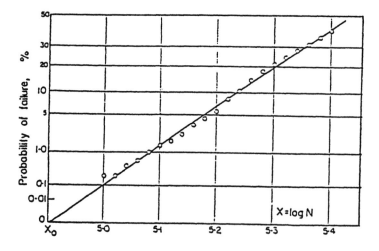

Figure 3.24 Probability *P* versus *N* in Weibull probability coordinates. (Weibull, W., see Reference 3. Copyright ARD NATO.)

imen fails or, if it was a run-out, the level is raised for the next test. The procedure is illustrated in Figure 3.25. Generally, stress increments are equal to about 5 percent of the initially estimated value, either upward or downward. In total, the procedure may comprise 15 to 30 tests.

Reliability of fatigue limit   For practical application there are standard engineering sources that list fatigue limit values S'_f. Since these are based on 50 percent probability, whenever the design requires safer limits the following fatigue limit can be used

$$S''_f = k_d S'_f \qquad [3.30]$$

where k_d is a reliability factor which accounts for the probability of the fatigue test data.

To illustrate, let us consider Figure 3.21. The standard fatigue $S-N$ curve is based on 50 percent reliability where the reliability is defined as

$$F = 1 - P \qquad [3.31]$$

Because of lower P, a curve plotted below, indicates a higher reliability. The lower $S-N$ curve in the figure shows $F = 0.9$ and $P = 0.1$ (10 percent probability of failure).

To calculate factor k_d for safer fatigue limit determination, consider the normal distribution of fatigue limit from Figure 3.21. As test results indicate,[12] the standard deviation σ of this distribution is equal to about 8 percent of the mean fatigue limit. A probability value $P = 0.1$ corresponds to a distance 1.282σ. Assume for simplicity that the $S-N$ diagram is plotted so that S is in linear and N in logarithmic scale, as per Figure 3.6(b). One can then state

$$S''_f = S'_f - 1.282\sigma = S'_f(1.0 - 1.282 \times 0.08) = 0.897\,S'_f \qquad [3.32]$$

It follows that for the specific case of 90 percent reliability, the k_d factor becomes

$$k_d = 0.897 \qquad [3.33]$$

In a similar way, for a 99 percent reliability we obtain the factor 0.814 which for a 50 percent reliability equals 1.0.

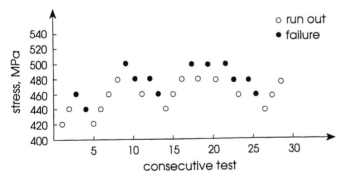

Figure 3.25 The staircase method.

3.2
ONE-DIMENSIONAL ANALYSIS OF MACHINE PARTS

We are now ready to relate the knowledge gained from testing the standard specimens to fatigue design analysis of actual machine parts. Since the geometry of many machine parts resembles simple bodies such as bars, beams and shafts, we start with a simplified stress analysis. Consider simple bodies subject to unidirectional stresses as follows: normal stress in bar under axial loading,

$$\sigma = \frac{P}{A} \qquad\qquad [3.34]$$

normal stress in straight beam in bending,

$$\sigma = \frac{My}{I} \qquad\qquad [3.35]$$

and shear stress in round shaft in torsion,

$$\tau = \frac{Tr}{J} \qquad\qquad [3.36]$$

The above expressions reflect two kinds of unidirectional stresses—normal and shear. When only one type of unidirectional stress is present, the stress analysis of a machine part is a one-dimensional analysis.

Consider machine parts with added geometric complexities, such as fillets, grooves, and holes. The irregularities in form render Equations (3.34) to (3.36) inaccurate because under load (as the tests showed) the stresses concentrate at the concavity. (See Figure 3.17.) Note that stress concentration is a critical factor for fluctuating loading, while for a monotonic (static) load its effects are not as material. A machine part with concavity will fail at essentially the same static load as the one without it. This paradox can be explained by the fact that a concavity does not influence the stress distribution at the time of static failure. When the material yields and plastic deformation takes place, stresses spread uniformly throughout the critical cross section and peaks are eliminated. At a time of failure the stresses reach the ultimate strength of material throughout the critical cross section.

The precise method to compute peak stresses in machine parts with concavities is to perform a multidimensional analysis based on the elasticity and plasticity theories. A shortcut, used in practice by machine designers, is to arrive at nominal stresses which are then multiplied by a corresponding notch factor K_f as follows

$$\sigma_{\text{peak}} = K_f \sigma_{\text{nom}} \qquad\qquad [3.37]$$

Notch factor K_f is defined by the equation

$$K_f = 1 + q(K_t - 1) \qquad\qquad [3.38]$$

where q is the notch sensitivity and K_t the theoretical stress concentration factor. (See

Section 3.1.1.) Values of K_t used in machine design are usually presented in the form of graphs. See Figures 3.26 through 3.31. These values are derived mainly with the help of mathematical analysis based on the theory of elasticity. (There are other known techniques for this analysis such as the photoelastic method and strain gauge method.)

3.2.1 Infinite fatigue life

Limit stress diagrams

For an infinite fatigue life the machine part must be designed to assure the working stresses due to loading are within the established limits. The limits, derived from fatigue test results, supplemented by data from monotonic tests, define a boundary of a safe domain. The data about cyclic and mean stresses are computed as a function of the fluctuating load and are plotted on a so-called limit stress diagram which serves as a practical design tool. Let us review the historical development of limit stress diagrams, which are in use today.

Gerber diagram The first limit stress diagram ever presented was by Gerber[13] in 1874. Originally the diagram presented S_{max} as a function of argument S_{min}. See Figure 3.32. A safe domain was bounded by a limit line in the form of a parabola through two limit points taken from test results—the ultimate tensile strength S_u and the fatigue limit S_f. The parabola is expressed by the equation

$$\frac{S_r}{2S_f} = 1 - \left(\frac{S_m}{S_u}\right)^2 \qquad [3.39]$$

where S_r is the stress range defined as

$$S_r = S_{max} - S_{min} = 2S_a \qquad [3.40]$$

A modified Gerber diagram that conforms to modern usage is shown in Figure 3.33. (See Reference 3) The diagram present S_r versus S_m based on the same parabolic law. As clarified by the modified approach, the parabolic limit line is symmetric in relation to tension and compression, as seen in the figure.

Though still applicable, the implementation of the parabolic law to the compression side of the diagram (to the left of line OA) is now known to be incorrect. According to tests there is no symmetry between tension and compression, that is, the negative and positive parts of the Gerber curve could not be symmetrical.

Goodman diagram A different diagram was established by Goodman[14] in 1899. In Goodman's original diagram the same variables S_{max} versus S_{min} are used, but his emphasis is on tension while the compression part is plotted nonsymmetrically (a more realistic approach). His limit stress diagram comprises a safe domain bounded by limit lines CB and DB. See Figure 3.34. Straight line CB connects two limit points taken from tests, S_u and S_f, and replaces the parabola of Gerber. Goodman line (CB) can be presented by the linear equation

$$\frac{S_r}{2S_f} = 1 - \left(\frac{S_m}{S_u}\right) \qquad [3.41]$$

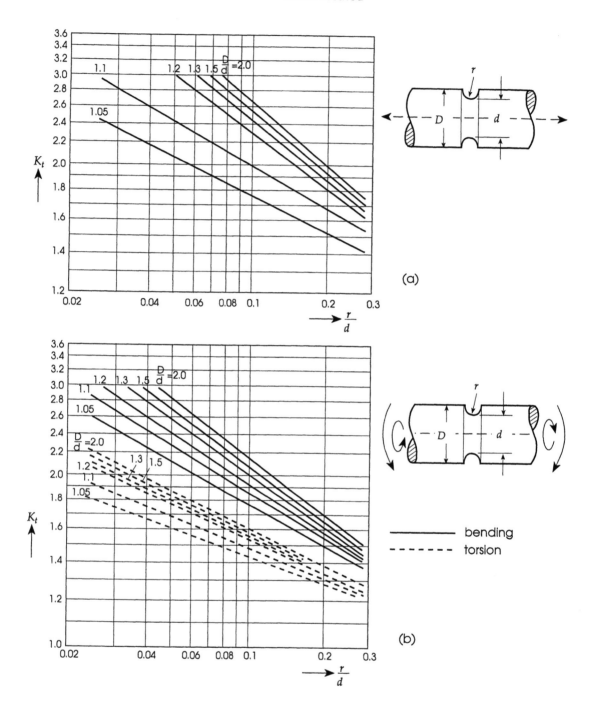

Figure 3.26 Stress concentration factors for a shaft with groove subject to axial load, bending and torsion. (Data from Peterson, R. E., see Reference 10.)

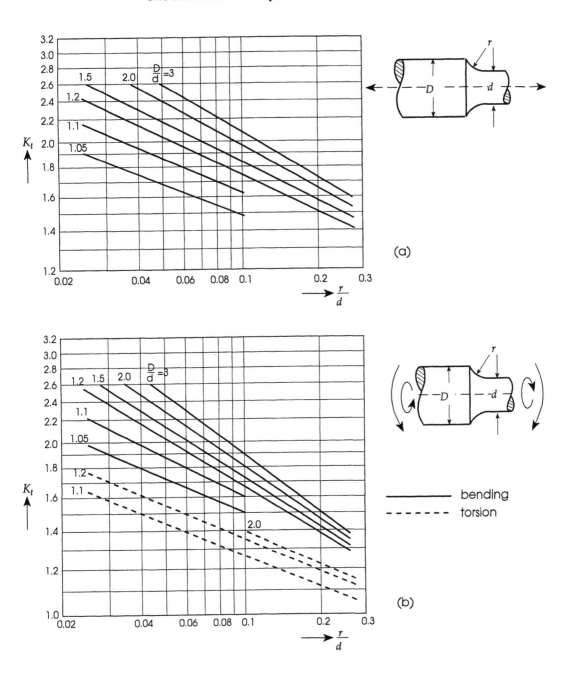

Figure 3.27 Stress concentration factors for a shaft with fillet subject to axial load, bending and torsion. (Data from Peterson, R. E., see Reference 10.)

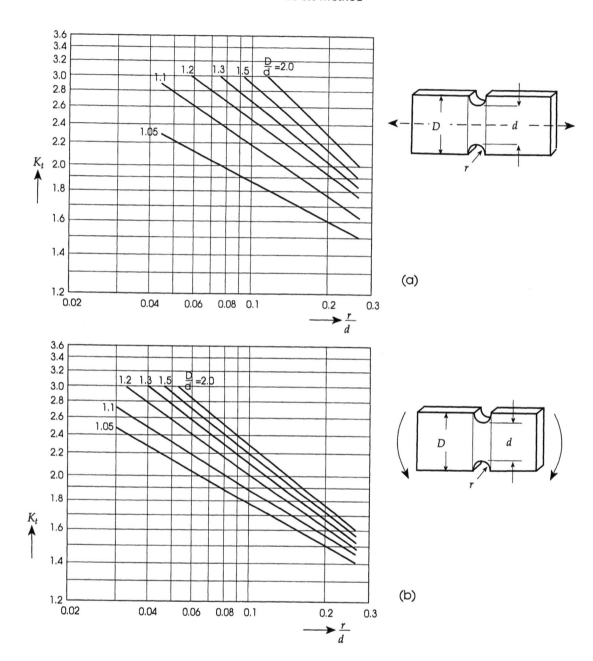

Figure 3.28 Stress concentration factors for a plate with
fillet subject to axial load and bending. (Data from
Peterson, R. E., see Reference 10.)

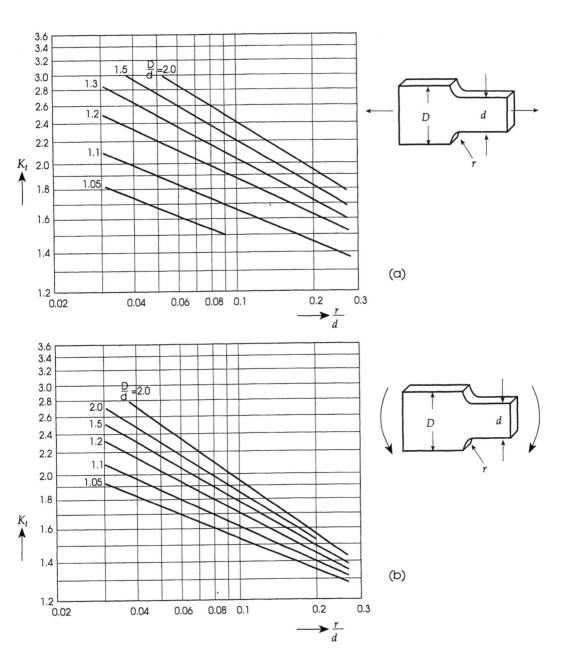

Figure 3.29 Stress concentration factors for a plate with fillet subject to axial load and bending. (Data from Peterson, R. E., see Reference 10.)

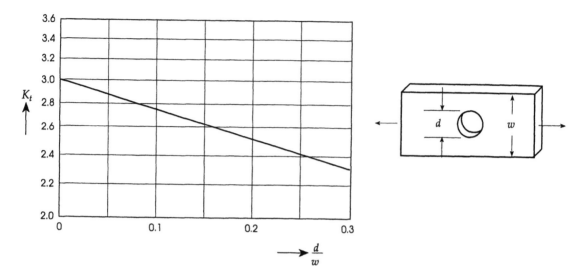

Figure 3.30 Stress concentration factors for a plate with a hole. (Data from Peterson, R. E., see Reference 10.)

(Goodman assumed that the maximum stress for infinite fatigue life should be half of the ultimate tensile strength when the minimum stress is zero which is a rough approximation.)

Figure 3.35 presents a modified Goodman diagram. Today, this tool is indispensable in design analysis. As modified from the original, it has the following basic features: (a) stresses S_{max} and S_{min} are presented as functions of S_m; (b) Goodman limit lines cross the ordinate axis at the points $\pm S_f$; (c) until now ignored, yield point S_y is considered.

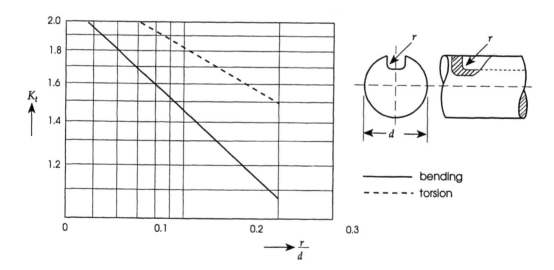

Figure 3.31 Stress concentration factors for a shaft with keyway in bending and torsion. (Data from Peterson, R. E., see Reference 10.)

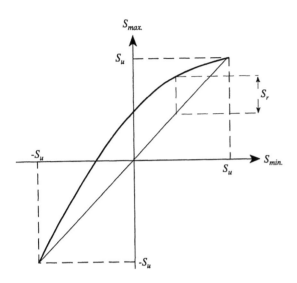

Figure 3.32 The Gerber diagram in original form.

The modified Goodman diagram considers the limits of plastic deformation defined as

$$S_{\max} \leq S_y \qquad\qquad [3.42]$$

Thus a new limit line *CD* is added to the diagram.

Consider Figure 3.36 showing several Goodman diagrams pertaining to shear stresses, for use in the design of helical springs.[15] Here $\tau_{\max}$ is presented as a function of $\tau_{\min}$, as in the original Goodman diagram. Different diagrams, favored on the European continent, are seen in Figure 3.37 where the limit lines *ABC* and *EDC* are drawn in such a way as to approximate Gerber's parabola.

Soderberg diagram In 1930, Soderberg[16] constructed a most conservative limit stress diagram presenting S_a as a function of argument S_m. As seen from Figure 3.38, the straight limit line AB connects, respectively, points $(0, S_f)$ and $(S_y, 0)$, fatigue limit S_f and yield point S_y as taken from test results. Using a straight line to connect the yield

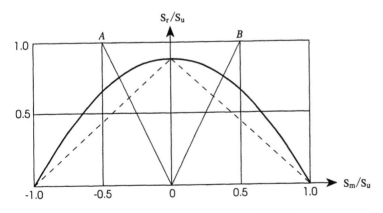

Figure 3.33 The Gerber diagram as used today.

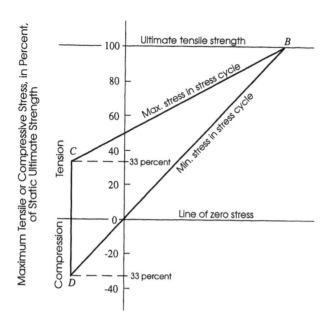

Figure 3.34 The Goodman diagram in original form.

point produces an overly conservative domain. Soderberg added later a limit line between points $(0, S_f)$ and $(-S_y, 0)$, symmetrical in respect to the vertical axis, rendering this part of the diagram to be incorrect. The Soderberg diagram was an effort with a specific purpose, that is, to create a design code for rotating shafts. (See Section 3.3.2.) Even though it offers a simple method for computing safety factors, in practice the conservative approach precludes its use.

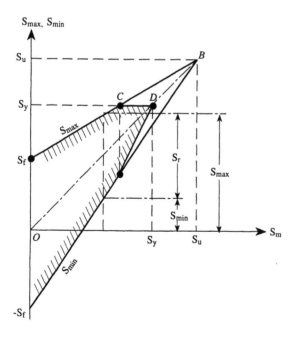

Figure 3.35 Modified Goodman diagram.

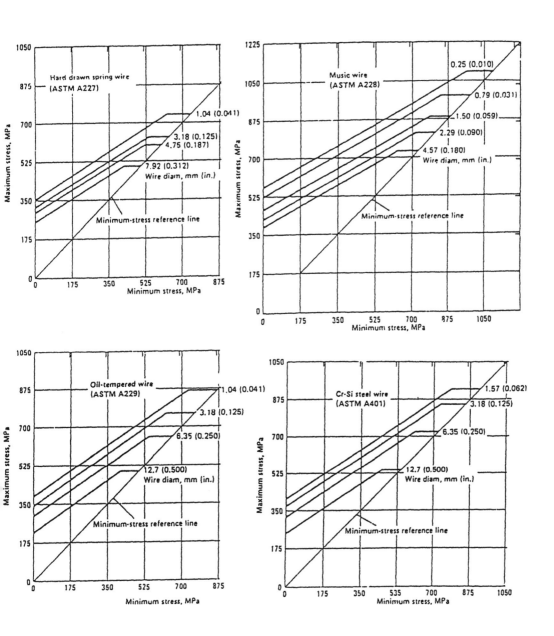

Figure 3.36 Goodman diagrams for ASTM steel wires for helical springs. Note: τ_{max} is plotted versus τ_{min}. ("Metals Handbook," see Reference 15. Reprinted by permission.)

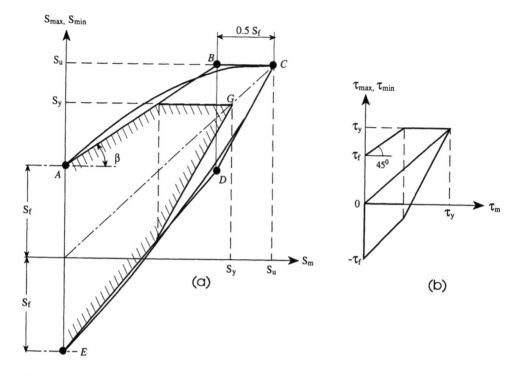

Figure 3.37 Approximate Goodman diagrams: (a) for axial loading; (b) for torsional loading of round bars.

Our discussion has been limited to a few chosen samples, taken from many others worthy of study. To present fully the advantages and disadvantages of each, we compare the methods described above. Figure 3.39 is a combined diagram with additional test data shown for comparison. (See Reference 8.) The superimposed limit lines of Gerber (positive side only), Goodman and Soderberg with yield line *CB*, confirm that the Soderberg line is overly conservative whereas the Gerber curve is hazardous. This leaves the combined line *ACB* (that is, Goodman line intercepted by the yield line) as the most

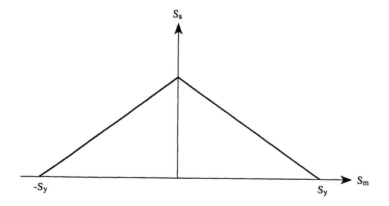

Figure 3.38 Soderberg diagram.

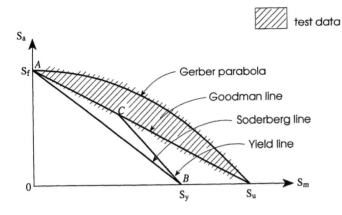

Figure 3.39 Comparison of different limit lines with test data.

suitable for design purposes. Note that line *ACB* is equivalent to line *ACD* of the modified Goodman diagram in Figure 3.35.

Having explained the validity of limit line in respect to tension, the compression domain has been purposefully omitted from our discussion. The above combined diagram (Figure 3.39) lacks the region of negative S_m. Experience shows that while a compressive load contributes to plastic deformation, it is the tensional load that causes a fatigue failure. Since we cannot disregard the plastic deformation occurring in a machine part, a definition of the yield line in compression must be provided for a practical application. A complete limit diagram which contains limits for both tensile and compressive stresses is shown in Figure 3.40. The right side of the diagram is comprised of a Goodman line combined with a yield line in tension. The left side shows the yield line in compression.

Let us also consider shear stresses in torsional loading. As known, positive and negative shear stresses are identical, limiting the need to the one side of the $\tau_a - \tau_m$ diagram. One can assume from testing unnotched cylindrical specimens that the fatigue limit in pure torsion is not influenced by mean stresses. (See Reference 8.) The limit line, therefore, first runs parallel to S_m axis and then follows the yield line as shown in Figure 3.41. The limit line may not be parallel to S_m for torsional specimen containing stress concentrations. This happens, for instance, in the case of coil springs. (See Figure 3.36.)

Application of fatigue notch factor   Using limit stress diagrams for parts with notches requires us to consider appropriate fatigue notch factors in order to allow for stress concentration. To illustrate the application of the fatigue notch factor, let us use

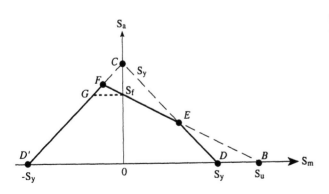

Figure 3.40 Complete limit diagram.

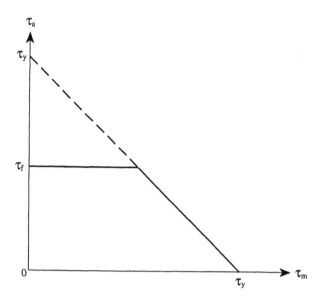

Figure 3.41 Limit stress
diagram for pure torsion
of a smooth specimen.

a standard limit diagram comprising a Goodman line in $S_a - S_m$ coordinates. See Figure 3.42. Concentrations of cyclic stresses caused by a notch may initiate a fatigue failure. To consider it, the K_f factor is introduced. Let us refer to point A', where both components, S_a and S_m, were multiplied by factor K_f. Since the local concentrations of steady stress do not contribute to failure, point A, obtained from multiplying S_a only, is more practical. S_m represents a static stress and we know that, when the material yields and plastic deformation takes place, static stresses spread uniformly throughout the critical cross section thus eliminating the need of a notch factor.

A more prevalent method of applying the fatigue notch factor is shown in Figure 3.43. This diagram presents a reduced fatigue limit, defined as S_f/K_f, and the Goodman

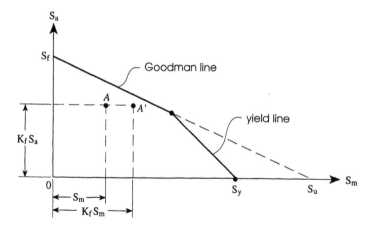

Figure 3.42 Working stresses in a limit diagram.
Working stresses are presented as nominal stresses multiplied by a stress concentration factor (in practical application only stress S_a is multiplied, indicated by point A).

line starts from the ordinate point S_f/K_f. Either of the two diagrams can be employed: Figure 3.42 or Figure 3.43. Point A in both diagrams reflects corresponding conditions. The coordinates in Figure 3.42 are $(S_m, K_f S_a)$, while in Figure 3.43 they become (S_m, S_a).

Factor of safety

The factor of safety for a given machine part at the most vulnerable location is the ratio of the strength of material to the working stress. Here the strength may be defined as the highest permissible stress that the working part can withstand during an infinite fatigue life. The safety factor is needed to support the known values against unknown values, that is, the properties of the part and the nature of the applied load. The greater the uncertainty, the larger the required safety factor. An increased safety factor, however, results in the excessive volume and weight of a machine. The importance of the safety factor is self-evident but, along with a presentation of different renderings that follow, it must be said that its choice in practice depends on the designer's experience. The factor may be as low as 1.1, when applied to some elements of aircraft design, or as high as 4.0, when applied to heavy stationary machinery.

Consider three cases illustrated in Figures 3.44, 3.45, and 3.46. In each case the limit is determined by the Goodman line combined with the yield line. Point A reflects the working condition and a notch factor is introduced using reduced fatigue limit S_f/K_f.

Case 1 First consider the case of a machine part where load vibrations are superimposed upon a steady working load. Here mean stress S_m is held at a closely guarded level, while cyclic stress amplitude S_a is subject to fluctuations due to load vibrations. Since the exact magnitude of S_a cannot be ascertained, steps must be taken to assure safety.

Let us represent the stress condition in a modified Goodman diagram with coordinates (S_{max}, S_m). See Figure 3.44. At the specified mean stress S_m the stress amplitude S_a equals AB, while the limit stress is CB. We introduce the factor of safety using the ratio

$$FS = \frac{CB}{CA} = \frac{S_{a,\text{lim}}}{S_a} \qquad [3.43]$$

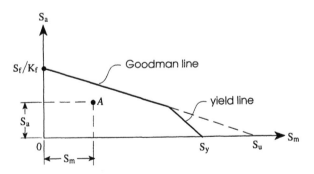

Figure 3.43 Working stresses in a limit diagram modified for stress concentration. Working stresses are presented as nominal stresses (indicated by point A).

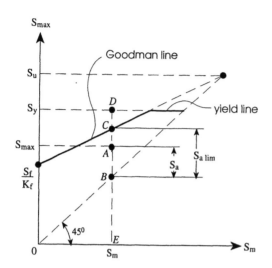

Figure 3.44 Factor of safety of a machine part when the mean stress is specified.

An additional problem to be considered is that of safeguarding the part against plastic deformation by keeping the maximum stress below the yield point (point A in the figure must be below point D). Hence we use a supplementary factor of safety

$$FS' = \frac{DE}{AE} = \frac{S_y}{S_{max}}$$ [3.44]

There are two safety factors to consider: Equation (3.43) and Equation (3.44).

Case 2   In the second case, we consider a bolted assembly where the bolts are tightened with a given force and at subsequent loading the load fluctuates at a constant amplitude. Here the minimum stress S_{min} is held at a guarded level, while the exact magnitude of stress range S_r cannot be ascertained.

The case is presented in Figure 3.45, which shows a Goodman line within coordinates S_{max} versus S_{min}. The factor of safety is introduced as

$$FS = \frac{CB}{CA} = \frac{S_{r,lim}}{S_r}$$ [3.45]

Here also there is a need for the second factor of safety FS', Equation (3.44), as in Case 1 above.

Case 3   The third case applies to the use of the safety factor in the analysis of machine design in general. Consider a stress condition where the magnitude of both mean stress S_m and stress amplitude S_a, cannot be ascertained. We assume that stress ratio, $A = S_a/S_m$, is strictly guarded.

The case is presented in Figure 3.46 which shows the Goodman line in coordinates (S_a, S_m). Lines OB, OB' and OB'' are locations of constant stress ratio A. The working condition is reflected in point A, while point B indicates the limit. Here the factor of safety becomes

$$FS = \frac{OB}{OA}$$ [3.46]

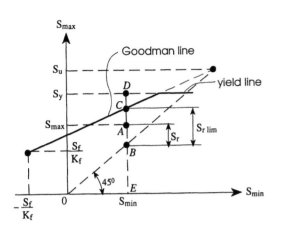

Figure 3.45 Factor of safety of a machine part when the minimum stress is specified.

Let us consider line *CD* parallel to Goodman line *FG*. It follows that

$$OD = \frac{S_u}{FS}$$

[3.47]

From geometrical similarity we get

$$\frac{S_a}{\frac{S_f}{FS}} = \frac{\frac{S_u}{FS} - S_m}{S_u}$$

[3.48]

from which follows

$$\frac{1}{FS} = \frac{S_m}{S_u} + K_f \frac{S_a}{S_f}$$

[3.49]

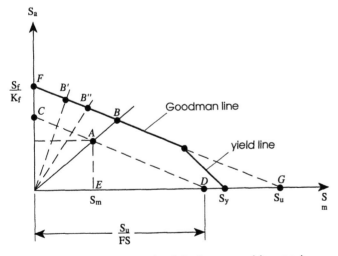

Figure 3.46 Factor of safety for a machine part when the stress ratio is specified.

Here also there is a need for a second safety factor FS'

$$FS' = \frac{S_y}{S_{max}}$$ [3.50]

Equation (3.49) may be applied to safety factors in practically all design cases, even though in its derivation the stress ratio A was assumed to be constant.

One single correlation can be used instead of the two above equations. See Figure 3.47. Returning to the Soderberg line, let us substitute in Equation (3.49) ultimate tensile strength S_u with yield point S_y. Consequently we obtain a new formula for a safety factor

$$\frac{1}{FS} = \frac{S_m}{S_y} + K_f \frac{S_a}{S_f}$$ [3.51]

which is known as the Soderberg equation. The reason only one equation is sufficient is because the Soderberg equation considers both the fatigue limit and the yield point. Because of its simplicity it is a convenient tool, but it may have limited practical application due to the overly conservative results.

3.2.2 Limited fatigue life

The main consideration in fatigue analysis is whether the part is to be designed for an infinite fatigue life or a limited life. The objective for the latter, the limited fatigue life, is to predict the number of cycles available within the fatigue life based on the magnitude of given stresses, or conversely, to determine the stresses based on a given number of cycles. For both objectives, a correlation between the number of cycles to failure and the magnitude of applied stresses becomes necessary. Two approaches are in use: graphical and analytical. The graphical method is based on life diagrams derived from tests, the analytical methods pertain to empirical models based on the test data.

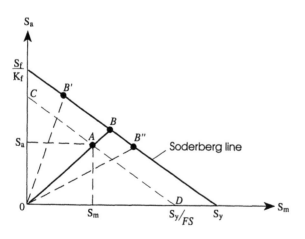

Figure 3.47 Factor of safety based on the Soderberg line.

Graphical approach

One approach to achieving a correlation between the number of cycles to failure and the magnitude of applied stresses uses a graphical representation of test results. The following derivation of the correlation is based on computation of constant life points plotted in a specially designed stress diagram.

To explain the creation of the diagram, let us consider S–N curves shown in Figure 3.10. In the figure, maximum stress S_{max} is plotted against fatigue life N and the individual curves correspond to different stress ratios R, the latter being defined by Equation (3.4). We now direct attention to a negative slope section of the S–N curves. By drawing vertical lines at various values of N, wherever they cross a S–N curve, values of S_{max} and R are obtained. From Equation (3.5) follows

$$\frac{S_a}{S_m} = \frac{1-R}{1+R}$$
[3.52]

Cross-plotting and connecting the constant life points on a new diagram, constant life lines in coordinates (S_a, S_m) are now apparent. See Figure 3.48. Starting at the point $(0, S_f)$, the lowest obtained line is for infinite fatigue life, while the connecting lines are for limited life.

The above diagram can have two applications: either with stresses S_a versus S_m, or with S_{max} versus S_{min}. This is explained schematically in Figure 3.49. The diagram in the figure is derived from the previous constant life diagram, rotating it around point 0 by 45 degrees and adding axes S_{max} and S_{min}. Adding the axes creates the possibility for presenting S_{max}, S_{min}, S_a and S_m, all in one diagram. This combination is shown in Figure 3.50 which presents standard stress diagrams used in design of machine parts in general.[17]

For illustration, other constant-life stress diagrams are shown in Figure 3.51. Instead of the stresses above, the diagrams show the plots of τ_{max} versus τ_{min} as is customary in design analysis of springs, specifically helical compression springs.

Figure 3.48 Constant life diagram.

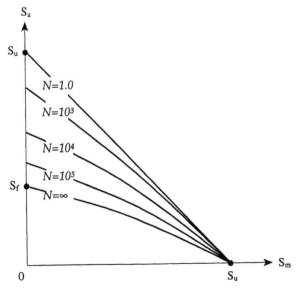

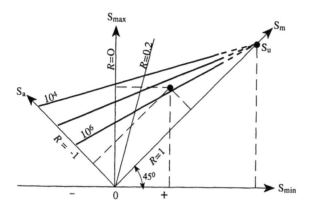

Figure 3.49 Schematic diagram of constant life showing S_{min}, S_{max}, S_a, and S_m.

Fatigue notch factor for limited fatigue life

The preceding discussion concerned the fatigue lives of smooth parts. When machine parts are endowed with notches or other geometrical irregularities, adding an appropriate fatigue notch factor becomes necessary. The fatigue notch factor in the limited life domain differs from that used in analysis for infinite life of similar parts. The analysis now must take into consideration yielding of the material, that is, stresses spread throughout the critical cross section smoothing out the uneven stress concentrations.

As we return to the $S-N$ curve in log-log coordinates, Figure 3.6(c), let us consider its negative slope but with an added stress concentration effect, as shown in Figure 3.52. The figure presents linear approximations of two $S-N$ curves for a machine part made of steel: one for an unnotched part and the other for a part with a notch. The effect of stress concentration on the latter part moves the $S-N$ curve downward because the stress concentration weakens the part. Along the lower $S-N$ curve (extending from $N = 10^6$ to infinity) the fatigue notch factor equals K_f and the fatigue limit is reduced to the value S_f/K_f. (K_f is derived for infinite fatigue life, see Section 3.2.1.) Tests show that for steels within the range of $N = 1$ to 10^3 the influence of notches is negligible. Therefore we assume that both curves overlap and the fatigue notch factor is 1.0. (Fuchs and Stephens[18] point to an interesting phenomenon that a notched specimen with a sharp groove may have a higher ultimate tensile strength than a uniform specimen due to the fact that the notched specimen, because of its geometrical form, will resist necking. This would make $K'_f < 1.0$ in the vicinity of $N = 1$.)

In the intermediate part of the lower $S-N$ curve, from $N = 10^3$ to 10^6, the fatigue notch factor (K'_f) continuously increases from 1.0 to its maximum value K_f, as expressed below,

$$1.0 \leq K'_f \leq K_f \qquad\qquad [3.53]$$

The following is a mathematical derivation of K'_f assuming $\log K'_f$ is a linear function of $\log N$ between points A and B,

$$\frac{0 - \log K'_f}{3 - \log N} = \frac{0 - \log K_f}{3 - 6} \qquad\qquad [3.54]$$

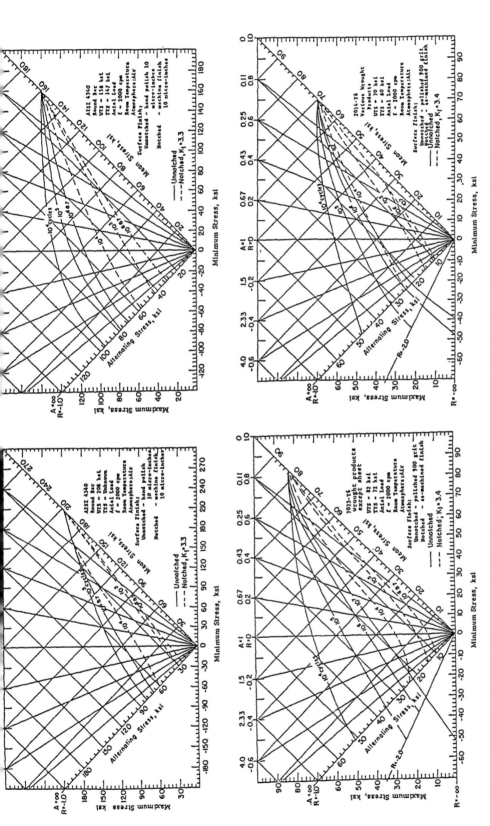

Figure 3.50 Constant life diagrams of different metals: (a) steel AISI 4340; (b) aluminum 7075-T6; (c) Titanium alloy Ti-6 Al4V. ("Military Handbook, Strength of Metal Aircraft Elements," see Reference 17.)

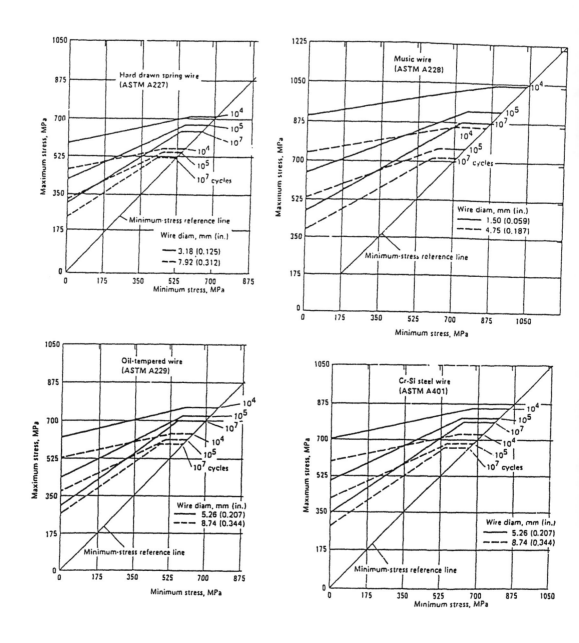

Figure 3.51 Constant life diagrams for ASTM steel wires for helical springs. Note: τ_{max} is plotted versus τ_{min}. ("Metals handbook," see Reference 15. Reprinted by permission.)

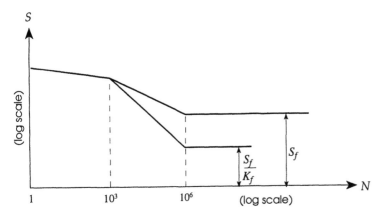

Figure 3.52 Effect of stress concentration on S-N curve.

From this, we obtain the fatigue notch factor for limited fatigue life

$$K'_f = \frac{1}{K_f} N^{\frac{\log K_f}{3}}$$

[3.55]

The same objective can be achieved for parts with notches using a constant life diagram with stress concentration effect, as seen from Figure 3.50. The basis of the diagram is taken from the analysis of infinite fatigue life where a reduced fatigue limit S_f/K_f is used instead of the ordinary S_f. (See Figure 3.43). In such a diagram only nominal stress values are considered (S_a, S_m, S_{max} and S_{min}), therefore the use of K'_f factor becomes unnecessary.

Analytical approach

The test data are an invaluable tool in machine design not only when it is done for specific cases but also when it represents a source for general use. In instances where specific fatigue data are missing, one can use empirical correlations derived from accumulated test data. These mathematical expressions can then take place of experimental data.

Constant-life stress correlation   This method is based on a linear approximation of constant life diagrams. Figure 3.53 presents a linear approximation of the $S-N$ curve in log-log coordinates. An exponential correlation between stress S and fatigue life N is obtained in the form

$$N = N_A \left(\frac{S}{S_A} \right)^{\frac{1}{a}}$$

[3.56]

Exponent a is the negative slope of the $S-N$ curve between A and B which equals

$$a = -\frac{\log S_B - \log S_A}{\log N_B - \log N_A}$$

[3.57]

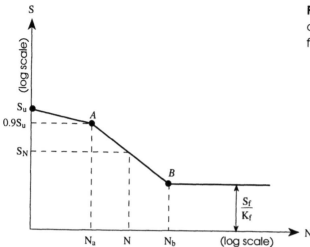

Figure 3.53 Approximate S-N diagram with stress concentration for design application.

For S_A we use the approximate value

$$S_A = 0.9 S_u \tag{3.58}$$

while for SB we have the expression

$$S_B = \frac{S_f}{K_f} \tag{3.59}$$

For steel, where $N_A = 10^3$ and $N_B = 10^6$, one gets from Equation (3.56)

$$N = 1000 \left(\frac{0.9 S_u}{S} \right)^{\log \frac{3}{0.9 S_u K_f}} \tag{3.60}$$

Equation (3.60) determines the number of cycles to failure N for a given S. A reciprocal of Equation (3.60) may be used when an opposite solution is required. Let us return to Equation (3.56) to derive its reciprocal

$$S = S_A \left(\frac{N}{N_A} \right)^a \tag{3.61}$$

which for steel parts becomes another useful formula

$$S = \frac{0.9 S_u}{\left(\dfrac{N}{1000} \right)^{\frac{1}{3} \log \left(\frac{0.9 S_u K_f}{S_f} \right)}} \tag{3.62}$$

Equations (3.60) and (3.62) form a linear approximation of the $S-N$ curve and are used for design purposes of steel parts. For metals other than steel, one can derive from Equation (3.56) equivalent formulas with corresponding values of N_A and N_B.

Similarly, the constant-life stress diagram may also be converted into a linear mathematical expression. The process is shown in Figure 3.54. In the right part of the figure a constant-life stress diagram is shown where the curves are replaced by straight lines. The lines cross the S_a axis at points C, C', C''. These points indicate values of fatigue strength, S_N, corresponding to the respective fatigue lives N. S_N may be considered a cyclic stress with zero mean, having the same effect on fatigue life as does the pair of stresses (S_a, S_m). As such, S_N is called the equivalent stress. Considering the similarity of triangles in the figure, we can state

$$S_N = S_u \frac{S_a}{S_u - S_m}$$

[3.63]

The left side of Figure 3.54 is a graphical representation of Equation (3.62), while the right side depicts Equation (3.63).

As can be seen from the figure, points C and G reflect equal fatigue life, thus supplementing each other. Life expectancy N can be traced from the known S_a and S_m following points D, C, G and H in succession. When stress values (S_a, S_m) are to be found, the same procedure is followed in reverse. The two equations together, Equations (3.62) and (3.63), allow a computation of life expectancy of a designed part when stresses S_a and S_m are known, or vice versa.

Berkovits universal method   This method computes fatigue life using a single correlation between the applied stresses and the life expectancy of a specimen.[19] By means of nondimensional parameters, the method succeeds in deriving analytically what standard constant life diagram shows graphically, that is, the lifetime damage induced

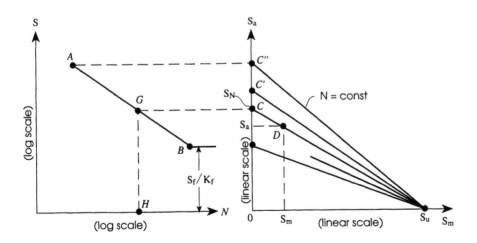

Figure 3.54 Linear approximation of a constant-life diagram.

by mean stress. A nondimensional parameter defining the fatigue life is introduced as follows:

$$\Omega = 1 - \frac{N}{N_{S_m=0}} \qquad\qquad\qquad [3.64]$$

where N denotes fatigue life at amplitude S_a and mean stress S_m, while $N_{S_m=0}$ is the fatigue life at amplitude S_a and zero mean stress. Condition $\Omega = 0$ is associated with fully reversed loading cycle, while $\Omega = 1$ denotes monotonic failure at the first cycle. Ω is a fatigue life parameter which reflects the effect of mean stress on the fatigue life. Thus, it is considered a function of stresses

$$\Omega = f(S_a, S_m, \xi_i) \qquad\qquad\qquad [3.65]$$

where ξ_i represents material constants. Further, a load function is introduced as a non-dimensional parameter

$$\phi = \frac{S_{max}}{S_u} \frac{1}{A} \qquad\qquad\qquad [3.66]$$

where A is the ratio of stress amplitude to mean stress as defined by Equation (3.3). Then fatigue life parameter Ω can be expressed as

$$\Omega = G\!\left(\phi\right) \qquad\qquad\qquad [3.67]$$

Based on experimental data, an empirical correlation is derived in the form

$$\Omega = 1 - \exp\!\left[-\frac{\phi^C}{B} \right] \qquad\qquad\qquad [3.68]$$

where B and C are material constants. Figure 3.55 is a graphical depiction of Equation (3.68). It renders a master curve which, in compact form, replaces the family of curves of a standard constant life diagram. This master curve or Equation (3.68) allow a computation of life expectancy of a designed part when stresses S_a and S_m are known or, vice versa, when life expectancy is known.

3.3
MULTIAXIAL ANALYSIS

Until now we have considered uncomplicated machine parts subject to simple loadings which caused unidirectional stresses, such as pure axial tension and compression, pure bending, or pure torsion. In this section we introduce the analysis of machine parts with

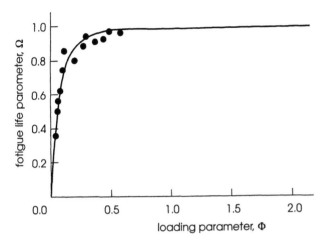

Figure 3.55 Master curve for effect of mean stress on fatigue life in Incoloy 901. (Data from Berkovits, A., and D. Fang, see Reference 19.)

complicated geometries and loadings, causing stress components to act simultaneously in different directions. Mathematical solutions are obtained by combining the stress components. Most components originate from the same source, that is, the machine's applied load. It is conservative to assume, therefore, that the components reach peak values, as well as valleys, simultaneously. In other words, they fluctuate in phase.

3.3.1 Combining fluctuating stresses

The theory of combining stress components was discussed in Chapter 2. There correlations were derived without giving due regard to stress fluctuation, because the problems there were limited to combining directional components only. We shall now consider other differences as well, that is, the cyclic and mean components. Before proceeding with the methods applied to combining all of the fluctuating components, let us reiterate one of the basic assumptions in the stress method: all derivations are based on elasticity theory and ignore plastic deformation.

The methods pertain to computing an equivalent cyclic stress amplitude S'_a and a equivalent mean stress S'_m by use of the appropriate combining formulas. Following are descriptions of several derivations, each based on a different approach.

von Mises equivalent stress

In this approach the equivalent stress amplitude and the equivalent mean stress are computed using von Mises correlation, Equation (2.49). Thus the respective equivalent stresses may be stated as follows. The equivalent stress amplitude equals

$$S'_a = \sqrt{\frac{1}{2}\left[\left(S_{a1} - S_{a2}\right)^2 + \left(S_{a2} - S_{a3}\right)^2 + \left(S_{a3} - S_{a1}\right)^2\right]} \qquad [3.69]$$

while the equivalent mean stress is

$$S'_m = \sqrt{\frac{1}{2}\left[\left(S_{m1} - S_{m2}\right)^2 + \left(S_{m2} - S_{m3}\right)^2 + \left(S_{m3} - S_{m1}\right)^2\right]}$$ [3.70]

S_{a1}, S_{a2} and S_{a3} are corresponding principal stress amplitudes and S_{m1}, S_{m2} and S_{m3} are corresponding principal mean stresses.

The resulting equivalent stresses S'_a and S'_m, even though representing a multiaxial fluctuating problem, are now considered as one-dimensional. Thus, we may proceed from here with a graphical or analytical approach (as presented in Section 3.2) and use it to consider an infinite fatigue life or a limited fatigue life, depending on the stress magnitude. Machine parts with stress concentration are discussed separately below.

Let us reexamine Equation (3.70) which has a basic fallacy. It disregards the known fact that a mean tension stress decreases fatigue life whereas a mean compressive stress extends it. This equation can be replaced as shown in the following approach.

Sines equivalent stress

The phenomenon omitted in Equation (3.70), namely, the different effect of mean tension versus mean compression in relation to fatigue life, was included in Sines' empirical expression[20] for equivalent mean stress S'_m. The following equation is simpler and gives better results

$$S'_m = a\left(S_{m1} + S_{m2} + S_{m3}\right)$$ [3.71]

where a is an empirical factor. The term in the brackets denotes equivalent hydrostatic pressure, an invariant that is independent of shear stresses. (See Chapter 2.) The equation gives expression to the experimental evidence, namely that mean shear stresses do not influence the fatigue life of a machine part.

Equation (3.71) was simplified by Fuchs and Stephens (Reference 18) as follows

$$S'_m = S_{m1} + S_{m2} + S_{m3}$$ [3.72]

The combined stress amplitude S'_a is computed as before by the von Mises equation, Equation (3.69). From here the fatigue problem is treated also as one-dimensional.

Stress concentration

We have deferred till now discussing the stress concentration effect. To consider the effect, we must introduce fatigue notch factors, as in the one-dimensional cases, discussed in Section 3.2. There a fatigue notch factor was applied to one-dimensional fluctuating stress. Now, because we are dealing with a number of stress components, we have to use an appropriate notch factor for each component.

Let us consider the equivalent stress amplitude by restating Equation (3.69) in a general form, where in lieu of the principal directions, global directions (x,y,z) are used as follows

$$S'_a = \sqrt{\frac{1}{2}\left[\left(S_{ax} - S_{ay}\right)^2 + \left(S_{ay} - S_{az}\right)^2 + \left(S_{az} - S_{ax}\right)^2 + 6(\tau^2_{axy} + \tau^2_{ayz} + \tau^2_{azx})\right]}$$

[3.73]

To consider stress concentration, each stress component must be multiplied by a respective fatigue notch factor, $K_{fx} S_{x,a}$, $K_{fy} S_{y,a}$, $K_{fz} S_{z,a}$, $K_{fxy} \tau_{xy,a}$, etc. In cases of infinite fatigue life the procedure is straightforward. The fatigue notch factors are constant as described in Section 3.2.1. For cases of limited fatigue life this is not true. When using fatigue notch factor K'_f: the fact that K'_f depends on N, renders the computation non-linear. See Equation (3.55).

A more accurate procedure for analyzing machine parts with stress concentration would be by using Equation (3.69), but only if it is possible to compute the exact stress distribution in the vulnerable locations. In that case, inserting the exact stress values renders the fatigue notch factors superfluous.

3.3.2 Design of rotating transmission shaft

An illustration of a multiaxial fatigue analysis is provided by the ASME standard code for designing transmission shafting.[21] It pertains to the design of a rotating shaft for infinite fatigue life. The method has its origin in Soderberg's work. (See Reference 16.) Contrary to the methods of von Mises and Sines, it takes into consideration mean shear stresses and uses Tresca theory for combined stresses.

Consider fluctuating stresses in the shaft. See Figure 3.56. In the application shown the stresses consist of mean and cyclic normal stresses and mean shear stresses. Mean normal stress $S_{x,m}$ is caused by steady axial force F,

$$S_{mx} = \frac{4F}{\pi d^2}$$

[3.74]

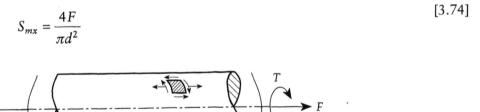

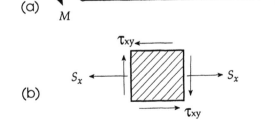

Figure 3.56 Rotating shaft:
(a) applied forces and moments; (b) stresses in shaft.

Cyclic normal stress $S_{x,a}$ is the result of rotating bending caused by moment M,

$$S_{ax} = \frac{32M}{\pi d^3} \qquad\qquad [3.75]$$

Mean shear stress $\tau_{xy,m}$ results from a steady torque T,

$$\tau_{mxy} = \frac{16T}{\pi d^3} \qquad\qquad [3.76]$$

An equivalent normal stress is defined by the equation

$$S'_{mx} = \frac{S_y}{FS'} = S_{mx} + K_f S_{ax} \frac{S_y}{S_f} \qquad\qquad [3.77]$$

where FS' is a relative safety factor according to Soderberg's Equation (3.51). Then the combined stress, based on the the Tresca criterion, Equation (2.54), equals

$$\tau_e = \sqrt{\left(\frac{S'_{mx}}{2}\right)^2 + \tau_{mxy}^2} = \sqrt{\frac{\left(S_{mx} + K_f S_{ax} \frac{S_y}{S_f}\right)^2}{4} + \tau_{mxy}^2} \qquad\qquad [3.78]$$

An overall safety factor FS is defined by

$$FS = \frac{S_y}{2\tau_e} \qquad\qquad [3.79]$$

Upon inserting Equations (3.74), (3.75) and (3.76) into (3.79), we obtain the expression of the safety factor

$$FS = \frac{\pi d^3}{32} \frac{S_y}{\sqrt{\left(\frac{Fd}{8} + BM\right)^2 + T^2}} \qquad\qquad [3.80]$$

where B is defined as follows

$$B = K_f \frac{S_y}{S_f} \qquad\qquad [3.81]$$

Equation (3.80) is known as the ASME equation for transmission shafting. It is as conservative as the original Soderberg diagram (Figure 3.38) and, in spite of the acknowledged conservative results, it is widely used in machine design.

3.4
CUMULATIVE DAMAGE

One of most difficult problems is to predict life expectancy in parts with spectrum loading, obviously applied where design is concerned with limited fatigue life. Spectrum loading (irregular loading, as the name suggests) represents a condition where the loading cycles vary and produce different magnitudes of peaks as well as valleys. See Figure 3.1. Until now our discussion has been restricted to problems where the maxima and minima of stresses do not change throughout the life with reference to one-dimensional and multiaxial stresses.

The fatigue life in spectrum loading is a function of accumulated effects on the part, at various stress levels, throughout its performance, that is, the function of cumulative damage. Though, a number of methods have been proposed to approach the analysis of cumulative damage, in practice one method is widely applicable, the Palmgren-Miner rule, described below.

3.4.1 Palmgren-Miner rule

The Palmgren-Miner rule is based on a crude assumption of a linear concept, that it is as if the changing of the sequence of nonuniform loading cycles did not affect the fatigue life. The rule is a combined result, introduced first by Palmgren[22] in analysis of ball bearings and adapted by Miner[23] for aircraft structures.

To describe the derivation of the Palmgren-Miner rule, let us replace the real sequence of cycles that takes place in irregular loading by an assumed sequence of blocks of uniform cycles. See Figure 3.57. Each block, comprising n_i number of uniform cycles, represents a corresponding load level i. At each load level the theoretical life

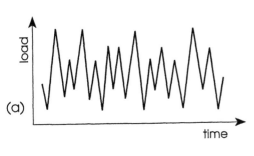

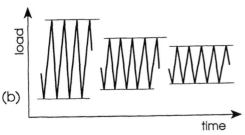

Figure 3.57 Referring to Palmgren-Miner rule: (a) spectrum loading; (b) blocks of uniform cycles.

expectancy is N_i number of cycles. The damaging effect of a single cycle at this level is assumed to be

$$D_i = \frac{1}{N_i} \qquad \text{[3.82]}$$

The damage after n_i cycles at this level is

$$n_i D_i = \frac{n_i}{N_i} \qquad \text{[3.83]}$$

We can express now the accumulated damage from the entire loading as

$$\sum n_i D_i = \sum \frac{n_i}{N_i} = \frac{n_i}{N_i} + \frac{n_2}{N_2} + \frac{n_3}{N_3} + \ldots \qquad \text{[3.84]}$$

One hundred percent damage of the machine part (which will cause fracture) can be expressed as

$$\sum n_i D_i = 1.0 \qquad \text{[3.85]}$$

This Equation (3.85), known as the Miner equation, forms the basis of the analysis. For practical application, the Miner equation can revised

$$\sum n_i D_i = C \qquad \text{[3.86]}$$

where C is an empirical constant which usually varies within the range

$$0.7 \leq C \leq 2.2 \qquad \text{[3.87]}$$

(Failures, however, have been known to occur at $C = 0.1$.) The total number of loading cycles at time of failure is

$$N = \sum n_i = n_1 + n_2 + n_3 + \ldots \qquad \text{[3.88]}$$

To facilitate the use of Equation (3.86), let us divide both sides by N, obtaining

$$\sum \frac{n_i D_i}{N} = \sum \frac{n_i}{N_i N} = \frac{C}{N} \qquad \text{[3.89]}$$

Let us introduce factors a_i, the number of cycles at load i relative to the total life time N (as explained in the next section), then

$$a_i = \frac{n_i}{N} \qquad \text{[3.90]}$$

It follows that

$$N = \frac{C}{\sum \dfrac{a_i}{N_i}}$$

[3.91]

The above expression is used to compute life expectancies of parts subjected to spectrum loading. Values of a_i are obtained by processing data from real loading histories, as described in Section 3.4.2. Even though in reality the damaging effect of a series of loading cycles is affected by the order, the derived method as expressed by Equation (3.91) is the most accessible to date.

3.4.2 Spectrum loading

To compute the life expectancy of a machine part using the method just described, factors a_i must be obtained. The factors identify, at each given load level, the number of cycles at a given level in relation to the overall number. The factors may be either estimated or taken from real loading histories. In the latter case the spectrum is measured with devices such as load cells and extensometers using strain gauges. Sometimes a strain gauge is attached to the part itself. Obviously, the data so obtained give more reliable results. Since monitoring machine parts in action requires investment in elaborate measuring and processing equipment, its application is limited mostly to such industries as aircraft, automotive and shipbuilding.

One or more representative spectrum data are recorded by continuous measuring of the strain or load during runs and processed into a finite number of blocks, each at a uniform load level. The two-step data processing is comprised of counting peaks and valleys, and creating a histogram with resulting factors a_i.

Counting peaks and valleys

The modern methods of counting the number of loading peaks and valleys consist of the selective recording of the more significant events. It is known that fatigue damage is caused mostly by a small number of extreme events. Several modern procedures are presented below where a range is the determining factor. A range is defined as the algebraic difference between successive valleys and peaks.

Rainflow method   This method[24] is a multi-step approach in counting peaks and valleys. Each series chooses for consideration different ranges, starting with extremes and at each iterative step decreasing the limits. The method has several variations. One of the variations is presented schematically. See Figure 3.58. A history of loading is simulated by a scenario where rain flows down a succession of pagoda roofs. The load-time history in the figure is rotated so that the time axis is directed vertically downward. In Figure 3.58 connections of peaks and valleys are represented by roofs. The assumptions for roofs and for rainflow are as follows:

- The flow starts at the highest point and continues to flow over the roof till it reaches the edge.
- It is assumed that if the roof is shorter than the one below, the rain drops verti-

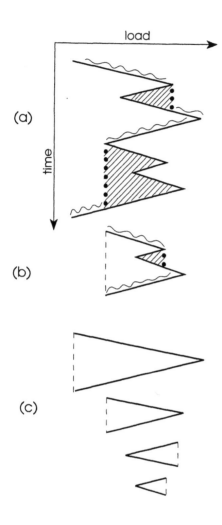

Figure 3.58 Rainflow method: (a) the rainflow path; (b) counting the omitted peaks and valleys; (c) selected cycles to be counted.

cally onto the one below; if the lower roof is shorter, the rain flows along the baseline, as if under, switching direction.

- The above is repeated for the remaining lesser ranges, stopping at a given range size.

The counted cycles are shown in Figure 3.58(c). (The rainflow method is especially applicable in fatigue analyses in connection with the strain method described in Chapter 4.)

Ordered overall range method   The method[25,26] pertains to defining, then selecting, the highest peaks and lowest valleys in a sequence. See Figure 3.59. If small ranges are present between the selected peaks and valleys, they are screened and discarded. The counted cycles are shown in Figure 3.59(b).

Racetrack method   Figure 3.60 illustrates the application of the method. (See Reference 26.) The history of events is simplified using an effective screening method to eliminate small ranges. The analogy here is a racetrack, the geometry of which has many right and left turns. In the figure the width of the racetrack is marked by w. At turns the racer cuts corners, moving along a straight unbroken path. Thus, smaller ranges are stretched forming a line. The recorded cycles are shown in Figure 3.60(b).

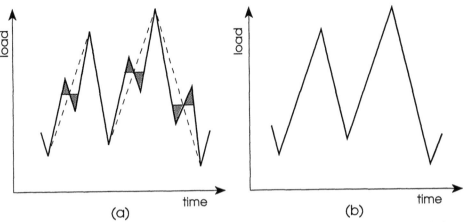

Figure 3.59 Overall-range method: (a) original history of events; (b) screened history.

Creating a histogram with resulting factors a$_i$

The computed numbers of events are now plotted into a cumulative distribution diagram as shown in Figure 3.61. The figure presents a continuous distribution. A histogram is fitted over it as an approximation. The histogram becomes, thus, the basis for determination of factors a_i.

To illustrate the computation of a_i, consider an eight-block histogram presenting eight load levels. As shown in Figure 3.61, the numbers of cycles in the progressive accumulation at respective levels equal

$$n_1 = 30,$$
$$n_1 + n_2 = 300,$$
$$n_1 + n_2 + n_3 = 3000,$$
$$n_1 + n_2 + n_3 + n_4 = 10,000,$$
$$n_1 + n_2 + n_3 + n_4 + n_5 = 85,000,$$
$$n_1 + n_2 + n_3 + n_4 + n_5 + n_6 = 300,000,$$
$$n_1 + n_2 + n_3 + n_4 + n_5 + n_6 + n_7 = 600,000,$$
$$n_1 + n_2 + n_3 + n_4 + n_5 + n_6 + n_7 + n_8 = 1,000,000$$

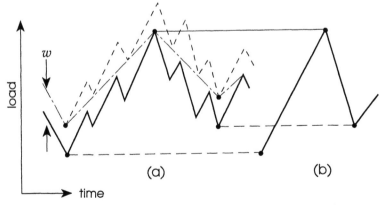

Figure 3.60 Racetrack method: (a) original history of events without a screening device; (b) screened history.

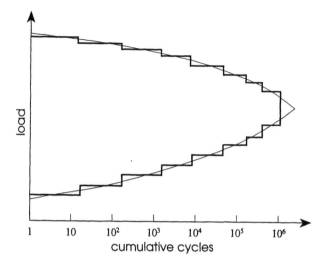

Figure 3.61 Cumulative load distribution.

The overall number of cycles is

$$N = \sum n_i = 1,000,000$$

Consequently, factors a_i, (identifying the numbers of cycles at each load level in relation to the overall cumulative number of cycles) are computed as follows

$$a_1 = \frac{30}{1,000,000} = 0.00003,$$

$$a_2 = \frac{270}{1,000,000} = 0.00027,$$

$$a_3 = \frac{2,700}{1,000,000} = 0.0027,$$

$$a_4 = \frac{7,000}{1,000,000} = 0.007,$$

$$a_5 = \frac{75,000}{1,000,000} = 0.075,$$

$$a_6 = \frac{215,000}{1,000,000} = 0.215,$$

$$a_7 = \frac{300,000}{1,000,000} = 0.3,$$

$$a_8 = \frac{400,000}{1,000,000} = 0.4$$

REFERENCES

1. Wohler, A. "Uber die Festigkeitsversuche mit Eisen und Stahl," *Z. Bauwesen,* 8 (1858) 641; 10 (1860) 583; 13 (1863) 233; 16 (1866) 67; 20 (1870) 73.
2. "Annual Book of ASTM Standards." 1980. Philadelphia, PA: American Society for Testing of Materials.
3. Weibull, W. 1961. *Fatigue Testing and Analysis of Results.* London: Pergamon Press.
4. Mischke, C. 1987. "Prediction of Stochastic Endurance Strength," *Trans. ASME, J. Vibration, Acoustics, Stress and Reliability in Design,* 109, 1, 113–122.
5. "ASME Handbook, Metals Engineering Design." 1965. Ed. Horger, O.J. New York: McGraw-Hill.
6. Faires, V.M. 1965. *Design of Machine Elements.* New York: Macmillan.
7. Heywood, R.B. 1962. *Designing Against Fatigue.* London: Chapman and Wall.
8. Juvinall, R.C. 1967. *Engineering Consideration of Stress, Strain and Strength.* New York: McGraw-Hill.
9. Grover, H.J. 1966. *Fatigue of Aircraft Structures.* Department of the Navy, NAVAIR 01–1A-13.
10. Peterson, R.E. 1974, *Stress Concentration Factors.* New York: John Wiley. 1953, *Stress Concentration Design Factors.* New York; John Wiley. 1951, "Design Factors for Stress Concentration," *Machine Design,* 23 (Feb.-July).
11. "A Guide for Fatigue Testing and the Statistical Analysis of Test Data." 1963. Philadelphia; American Society for Testing of Materials. ASTM STP-91-A.
12. Mischke, C.R. 1974. "A Rationale for Mechanical Design to a Reliability Specification." ASME Design Engineering Technical Conference, New York.
13. Gerber, W. 1874. "Bestimmung der Zulassigen Spannungen in Eisenkonstruk-tionen." *Z. Bayer, Arch, Ing. Ver.,* 101.
14. Goodman, J. 1899. *Mechanics Applied to Engineering.* London: Longmans.
15. "Metals Handbook." 1990. 10th ed. Metals Park, OH: ASM International, vol. 1.
16. Soderberg, C.R. 1930, "Factor of Safety and Working Stress." *Trans. ASME,* 52, part 1, APM 52–2. 1933, "Working stresses," *Trans. ASME,* 55, part 2, APM 55–16.
17. "Military Handbook, Strength of Metal Aircraft Elements." 1975. MIL-HBDK-5, U.S. Department of Defense, Washington, D.C.
18. Fuchs, H.O. and R.I. Stephens. 1980. *Metal Fatigue in Engineering.* New York: John Wiley.
19. Berkovits, A., and D. Fang. 1993. "An Analytical Master Curve for Goodman Diagram Data." *Int. J. Fatigue,* 15, no. 3, 173–180.
20. Sines, G. 1959. "Behaviour of Metals under Complex Static and Alternating Stresses," in *Metal Fatigue.* Ed. Sines, G. and G.L. Waisman. New York: McGraw-Hill, 197.
21. "Design of Transmission Shafting." 1985. ANSI/ASME, B106.1M.
22. Palmgren, A. 1924. "Die Lebensdauer von Kugellagern." *Zeitschrift VDI,* 68, 14, 339–341.
23. Miner, M.A. 1945. "Cumulative Damage in Fatigue." *Trans. ASME, J. Appl. Mech.,* 67, A159.
24. Matsuishi, M., and T. Endo. 1968. "Fatigue of Metals under Varying Stresses." paper presented to Japanese Soc. Mech. Eng., Fukuoka, Japan, March.
25. Hayes, J.E. 1965. "Fatigue Analysis and Fail-Safe Design." in *Analysis and Design of Flight Vehicle Structures,* ed. Bruhn, E.F. Cincinnati, OH: Tri-State Offset Co., Cincinnati, OH, 1965, C13–1 to C13–42.
26. Fuchs, H.O., D.V. Nelson, M.A. Burke and T.L. Toomay. 1977. "Shortcuts in Cumulative Damage Analysis." in *Fatigue under Complex Loading,* ed. Wetzel, R.M. Warrendale, PA: SAE.

4

STRAIN
METHOD

The strain method of fatigue analysis (also known as "low-cycle fatigue" method) predicts the life expectancy of a machine part based on the magnitude of fluctuating strain. The method has its beginning in the early 1950s in observations of metals endowed with low cycle fatigue. Low cycle fatigue (LCF) is caused by high, fluctuating stresses and strains which result in a life span ranging from 1 to 10,000 cycles, as opposed to high cycle fatigue (HCF) where the life span is more than 10,000 cycles. Today, the strain method is in frequent use by machine designers in LCF and HCF analyses and it is recommended by ASME,[1] SAE[2] and ASTM.[3]

The following is a two-part theory presentation of the strain method: (a) basic one-dimensional theory applicable to simple bodies subject to axial loading; (b) multiaxial theory considering the true geometry and true loading. Throughout this chapter the loading is assumed to be periodically and uniformly repeated; in instances when random loading is considered, the methods presented in Chapter 3 (Section 3.4) are applicable.

4.1
ONE-DIMENSIONAL ANALYSIS

The empirical information that leads to the theory of the strain method is obtained from low cycle fatigue tests where a metal specimen is subjected to cyclic loading. Before proceeding with the cyclic loading, a less complex case of the monotonic loading must be understood. Even though the monotonic properties are different from the cyclic properties, an understanding of the behavior is essential to formulating the theory of the strain method.

4.1.1 Monotonic properties (tension test)

Consider a standard tensile test where a metal specimen with a constant cross section is subjected to a monotonously increasing tensile force. See Figure 4.1. During the test the changing force and elongation are observed and recorded for stress and strain computation. For design applications, the axial stress is computed using the equation

$$S = \frac{P}{A_0} \qquad\qquad\qquad\qquad [4.1]$$

where P is the load and A_0 is the area of the original cross section (that is, assuming the cross section to remain unchanged). The strain is computed using the equation

$$e = \frac{l - l_0}{l_0} \qquad\qquad\qquad\qquad [4.2]$$

where l_0 is the original length, l is the length under load P, and S and e are called engineering stress and engineering strain, respectively. Figure 4.2 shows plots of S versus e (so-called engineering stress-strain curves) for two different steels as obtained from tensile tests.

Under the monotonously increasing loading the specimen undergoes first an elastic deformation which is reversible and, after the load increases beyond the yield point, a plastic deformation takes place leaving a permanent distortion in the geometry. Figure 4.3 shows the stress and strain behavior in the specimens under the increasing load as applied to a high carbon steel. Point P marks the proportional limit: up to this point the stress-strain correlation is linear. Point P is followed by the yield point Y. If the load is removed before reaching point P the specimen bears no change in its geometry. Beyond point Y a plastic deformation occurs. For steels it is an accepted practice to place

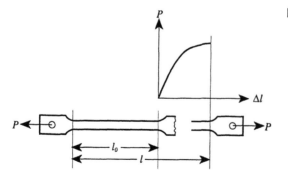

Figure 4.1 Tension test.

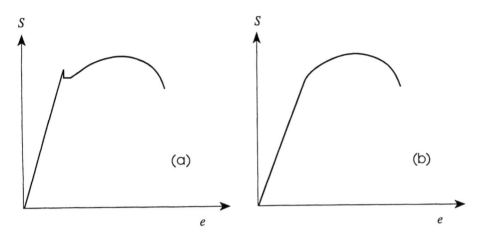

Figure 4.2 Stress-strrain curves for (a) mild steel and (b) high-carbon steel.

point Y at 0.2% offset in relation to proportional limit P (as seen in the figure). It should be noted that in design applications point P is assumed to coincide with point Y, as if the stress strain correlation is linear throughout.

True stress and strain

Before plastic deformation takes place, the engineering stress and strain defined by Equations (4.1) and (4.2) are applicable for lower loads. A more functional definition becomes necessary once plastic deformation is present. Let us improve the accuracy of Equations (4.1) and (4.2) as applied to the plastic domain by taking into consideration the change in cross-section area A. As the load P increases the cross-section area A decreases. As a rule, the volume of the specimen during the plastic domain is assumed to be constant. Thus,

$$Al = A_0 l_0 \qquad\qquad [4.3]$$

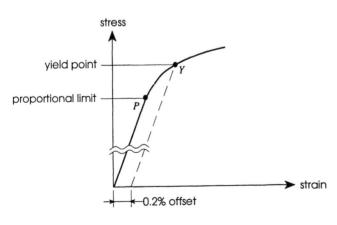

Figure 4.3 Definition of proportional limit and yield point as applied to high carbon steel.

Strain Method

Combining the latter with Equation (4.2) we get

$$A = \frac{A_0}{1 + e} \qquad [4.4]$$

We now introduce a stress σ, sensitive to the changes of A, referred to as true stress,

$$\sigma = \frac{P}{A} \qquad [4.5]$$

Comparing the two definitions, Equations (4.1) and (4.5), we find that

$$\sigma = \frac{P}{A_0}\frac{A_0}{A} = S(1 + e) \qquad [4.6]$$

(Note: in the elastic range, where e is very small, the right hand factor in the above equation approaches 1 and σ approaches S.)

To introduce the true strain now, we first present it in its differential form

$$d\varepsilon = \frac{dl}{l} \qquad [4.7]$$

The true strain upon integration equals

$$\varepsilon = \int_{l_o}^{l} \frac{dl}{l} = \ln\frac{l}{l_0} = \ln(1 + e) \qquad [4.8]$$

Thus we have obtained the expressions for the true stress and strain and for the engineering stress and strain. To demonstrate graphically the differences, let us refer to Figure 4.4 where two curves, applicable to a high carbon steel, are compared: true stress-strain curve (OF) and engineering curve (OD). In the figure, point O designates the start of the loading process and point Y the yield point. Within the elastic domain (OY) the difference between cross section A and A_0 is negligible and both curves merge into a single line. Within the plastic domain, continuous effects of the increasing load are

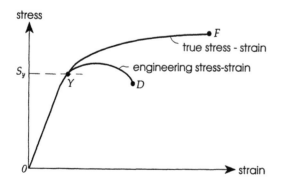

Figure 4.4 Comparing engineering stress-strain curve and true stress-strain curve applicable to a high carbon steel.

reflected by the diverging curves, YD and YF. Failure of the specimen is marked by points D and F, respectively.

Monotonic stress-strain correlation

Let us observe the stress and strain behavior in the monotonic test. Figure 4.5 shows the loading path OA in true coordinates (ε,σ). Up to the yield point the stress σ rises in a direct proportion to strain ε following the Hooke's law of the elasticity theory

$$\sigma = E\varepsilon \qquad [4.9]$$

In the plastic domain beyond the yield point the true stress-strain relationship is expressed by Hollomon correlation[4]

$$\sigma = K\varepsilon^n \qquad [4.10]$$

where K is a strength coefficient and n a strain hardening exponent. At point A the total strain ε comprises two components, an elastic strain and a plastic one. This is expressed by

$$\varepsilon = \varepsilon_e + \varepsilon_p \qquad [4.11]$$

As seen in the figure, plastic strain ε_p corresponds to a permanent plastic deformation that remains after the specimen is unloaded and the stress is zero.

The elastic and plastic components are defined as follows. The definiton of elastic strain ε_e follows from Equation (4.9),

$$\varepsilon_e = \frac{\sigma}{E} \qquad [4.12]$$

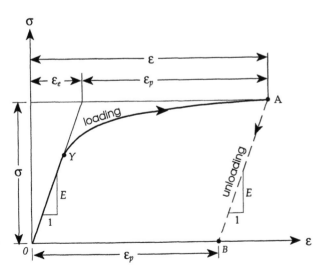

Figure 4.5 Loading path beyond the yield point in monotonic test.

The definition of plastic strain ε_p is derived from Equation (4.10). This correlation is based on experimental data of Hollomon, as shown in Figure 4.6. (See Reference 4.) Because the experiments were performed within a very limited range (strains varying from 0.01 to 0.6), the difference between the total strain and plastic strain was negligible. Consequently one can express ε_p with sufficient accuracy using the following equation

$$\varepsilon_p = \left(\frac{\sigma}{K}\right)^{\frac{1}{n}}$$

[4.13]

In addition, we assume that Equation (4.13) applies not only to the larger strains used by Hollomon but also can be extrapolated to strains of less than 0.01, aproaching the yield point. Thus the total strain ε can be expressed in the form

[4.14]

$$\varepsilon = \frac{\sigma}{E} + \left(\frac{\sigma}{K}\right)^{\frac{1}{n}}$$

4.1.2 Cyclic properties

Let us now relate to a fatigue state where cyclic rather than monotonic loading prevails. Instead of a tensile test considered above, we will introduce here the so-called low cycle fatigue tests, where the behavior of a cyclically loaded specimen is monitored. The test machinery and its control equipment are more complicated, like that widely used in the automotive and aircraft industries. (See Reference 3.)

Let us consider a typical electrohydraulic test system as shown in Figure 4.7, similar to the one described in Chapter 3 (Section 3.1). The system is fully automated: It comprises a computer (d) for test programming and control, a servo-controller (e) to

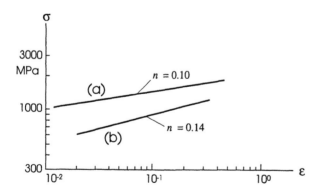

Figure 4.6 Monotonic stress-strain curves in log-log coordinates for (a) steel SAE 1055 WQT1000 and (b) steel SAE 1055 WQT1300. (Data from Hollomon, J. H., see Reference 4.)

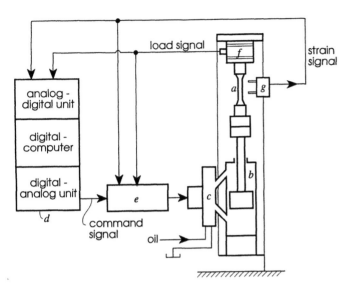

Figure 4.7 A computerized electrohydraulic low-cycle fatigue test system: (a) specimen, (b) hydraulic actuator, (c) servo-valve, (d) computer, (e) servo-controller, (f) load cell, (g) extensometer.

monitor the applied load, a load cell (f) and an extensometer (g) to check the specimen's changing geometry. See Figure 4.8. The computer reads the load and strain signals from the sensors and responds with command signals to the controller. For each load fluctuation there is a signal defining the load magnitude and direction in accordance with a programmed load history in the computer which includes the consecutive maxima and minima of force, stress, displacement and strain. The original signals are analog and are translated into digital for computer input.

To observe the behavior of a cyclically loaded specimen, we shall distinguish between two types of tests. The tests with stress control have the prescribed stress fluctuation data stored in computer memory, while the tests with strain control have the strain data stored. The controlling factor can be either constant or purposely varying.

Let us consider a stress controlled test first. Imagine a metal specimen having hypothetical properties that stay constant under load cycling. The specimen is subjected to completely reversed loading, symmetric tension and compression beyond the yield point, following one after the other in succession. The process is presented schematically in Figure 4.9(a). The loading history begins with tension at point O. At first, the deformation is elastic and is represented by straight line OA. Beyond point A, the yield point in tension, the deformation is plastic, as reflected by curve AB. Beyond point B, the turning point, it follows the unloading, as indicated by BO'. Since it is assumed that no changes in metal properties take place, the subsequent reverse loading (compression) has an equal but symmetrically opposite pattern. From O' to A' the deformation is again elastic and from A' to B' it is plastic. Point B' is again a turning point, after which the unloading brings us back to the original point O. This pattern goes on, resulting in a stable hysteresis loop.

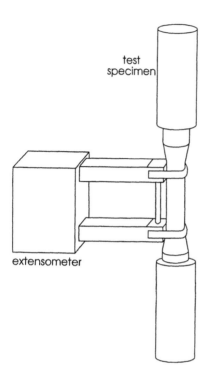

Figure 4.8 Test specimen with attached extensometer for longitudinal strain measurement.

test specimen

extensometer

In real life, the metals do respond to the cyclic loading with changes of either strain hardening or strain softening (known as cycle hardening or cycle softening). Consider again the same test as above, using a metal specimen with a cycle hardening response. The process is presented schematically in Figure 4.9(b) showing continuously changing hysteresis loops. We note in the figure that, starting with the first reverse loading (compression), the process is different than the one shown in Figure 4.9(a), as reflected by

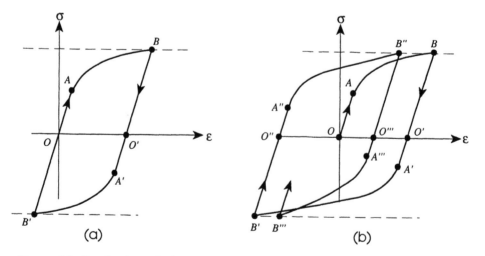

Figure 4.9 Cyclic stress-strain curves for (a) assuming that strain cycling does not affect material properties and (b) taking into consideration material response due to Bauschinger effect and strain-cycle hardening.

path O′A′B′. The changed pattern continues with another unloading B′O″, after which the specimen is reloaded again, as indicated by O″A″B″, and so forth. The start loading path, OAB, and the first reverse loading path, O′A′B′, are nonsymmetrical: yield point A′ is less than yield point A. This is known as the Bauschinger effect.[5] The cycling process produces strain hardening evident in the changing positions of turning points B, B′, B″, and so on. While the cyclic hardening prevails, the Bauschinger effect diminishes.

To show both responses to cyclic loading, cycle hardening and cycle softening, let us describe the behavior in stress controlled tests of two specimens of different metals. See Figure 4.10. For both specimens the stress σ has a prescribed constant amplitude. Part (a) of the figure reflects the behavior of a cycle hardening response indicated by the fact that the strain ε follows with an asymptotically decreasing amplitude. After a large number of cycles a stable state is achieved as presented in the figure by the stabilized hysteresis loop. Part (b) of the figure shows a cycle softening response. On the other hand, here the strain ε follows with an asymtotically increasing amplitude, reaching at the end a final hysteresis loop.

One observes equivalent phenomena in strain controlled tests. Let us describe again two responses where the strain ε has a prescribed constant amplitude. See Figure 4.11. Part (a) of the figure shows a cycle hardening response where the stress σ follows with an increasing amplitude till the stable hysteresis loop is reached. Part (b) of the figure shows a cycle softening response where the stress amplitude decreases toward the hysteresis loop.

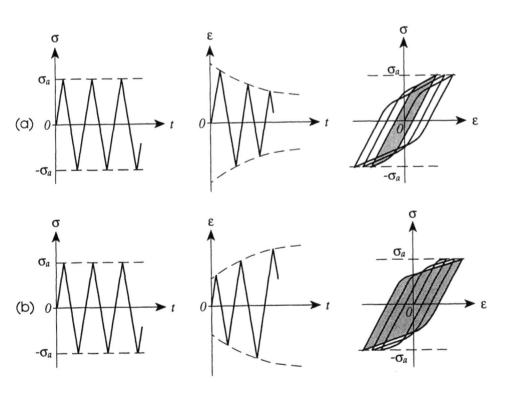

Figure 4.10 Stress versus strain in a stress controlled test for (a) specimen made of cycle-hardening material and (b) specimen made of cycle-softening material.

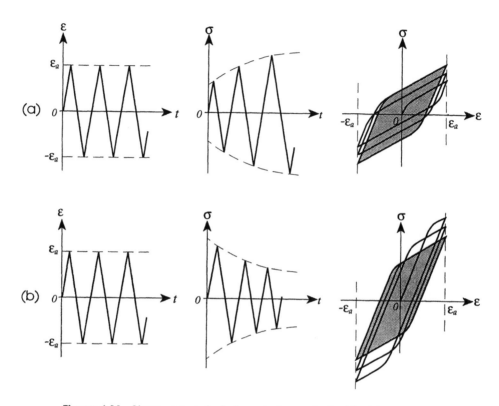

Figure 4.11 Stress versus strain in a strain controlled test for
(a) specimen made of cycle-hardening material and (b) specimen made
of cycle-softening material.

The purpose of the above tests is to produce a stabilized hysteresis loop as shown in Figure 4.12. The size of the hysteresis is defined by the stress range $\Delta\sigma$ and the strain range $\Delta\varepsilon$. Both ranges are functions of the imposed fluctuating load. Because of the fact that the stress control at large loads is cumbersome, the strain control tests are more convenient even though both tests give similar results.

Cyclic stress-strain correlation

Let us examine the stress and strain amplitudes in the stable state, $\Delta\sigma/2$ and $\Delta\varepsilon/2$. See Figure 4.12. There are several ways of arriving at a correlation of the two amplitudes. We shall describe two: a multispecimen and a multistep test program. In the former, a series of same metal specimens are tested. In the latter, one specimen undergoes a succession of tests.

In the multispecimen test program the specimens are subjected to load fluctuations under constant strain control. The specimens are cycled, each one at a different strain amplitude, until their condition is stabilized. See Figure 4.13. A series of resulting hysteresis loops is plotted in a common $\sigma-\varepsilon$ diagram. Connecting the hysteresis tips we get a stress-strain curve. See Figure 4.14. The curve represents a correlation between the cyclic stress and strain amplitudes.

In the multistep test program (incremental step program), a specimen is subjected to load successions of programmed blocks comprised of gradual increases and decreases

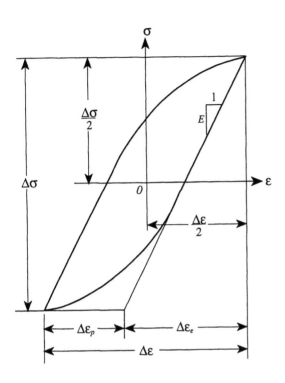

Figure 4.12 Stabilized hysteresis loop showing stress and strain amplitudes in the stable state.

of strain amplitudes. See Figure 4.15. The resulting series of hysteresis loops provide a cyclic stress-strain curve as before. Comparison shows that the stress-strain curves from the two types of tests[6] differ somewhat. See Figure 4.16.

Let us now compare the form of the cyclic stress-strain curve with a monotonic one. The difference can be noted in Figure 4.17 which shows both monotonic and cyclic stress-strain curves of two metals: (a) steel SAE 4142 and (b) aluminum 2024-T4.[7] The curves overlap in the elastic range and diverge in the plastic range. The plastic range of the cyclic curve in part (a) is below the plastic range of the monotonic curve. This is due to the cycle softening response of the given specimen. An opposite effect is noted in part (b) which is due to the cycle hardening response of that specimen. It should be noted that the metal production process has a direct bearing on the cyclic response. Annealed metals, free of any prior treatment, stressed in plastic range and later released, tend to be cycle hardening. On the other hand, treated metals, subject to residual

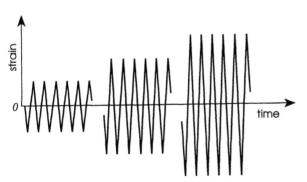

Figure 4.13 Multispecimen test program.

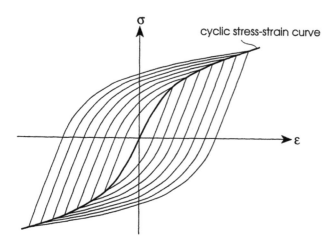

Figure 4.14 Derivation of the cyclic stress-strain curve.

stresses released during the cyclic loading, tend to be cycle softening. Figure 4.18 illustrates different behaviors of OFHC copper: annealed copper displays cycle hardening and cold-worked copper shows cycle softening.[8]

Mathematical derivation   After the graphic presentation of the cyclic stress-strain curve let us derive the stress-strain correlation in mathematical form for the use in machine design. To do this, let us compare the curves in Figures 4.5 and 4.16 and note that the cyclic and the monotonic curves have similar forms. The similarity in graphical form enables us to use the mathematical expression of the monotonic correlation as the basis to obtain the cyclic correlation, as shown below.

By definition, the cyclic strain amplitude comprises two components, elastic and plastic strain amplitudes

$$\frac{\Delta \varepsilon}{2} = \frac{\Delta \varepsilon_e}{2} + \frac{\Delta \varepsilon_p}{2} \qquad [4.15]$$

By analogy with Equation (4.12) for monotonic loading, the elastic strain amplitude in cyclic loading can be expressed as

$$\frac{\Delta \varepsilon_e}{2} = \frac{\Delta \sigma}{2E} \qquad [4.16]$$

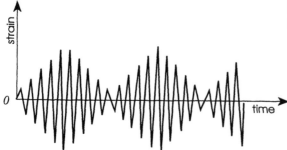

Figure 4.15 Multistep (incremental step) test program.

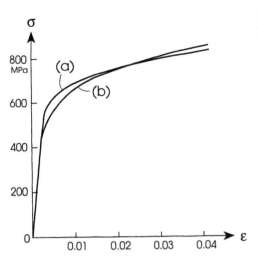

Figure 4.16 Cyclic stress-strain curves for RQC-100 steel: (a) from multispecimen test; (b) from multistep test. (Data from Tucker, L., and S. Bussa, Reference 6.)

The plastic strain amplitude in cyclic loading, analogous to Equation (4.13) for monotonic loading, equals

$$\frac{\Delta \varepsilon_p}{2} = \left(\frac{\Delta \sigma}{2K'} \right)^{\frac{1}{n'}}$$

[4.17]

Upon summation, Equations (4.16) and (4.17) provide the total cyclic strain amplitude

$$\frac{\Delta \varepsilon}{2} = \frac{\Delta \sigma}{2E} + \left(\frac{\Delta \sigma}{2K'} \right)^{\frac{1}{n'}}$$

[4.18]

Thus we arrived at the cyclic stress-strain correlation which is similar to Equation (4.14) for monotonic loading. K' is the cyclic strength coefficient and n' is the cyclic strain hardening exponent. Figure 4.19 exhibits cyclic test results,[9] confirmed using Equation (4.18).

Equation (4.18) may be simplified with the help of data from cyclic tests, that is, the stress and strain that cause a specimen's fracture at the first reversal of cyclic loading. The stress, referred to as fatigue strength coefficient, is denoted by σ'_f. The strain, referred to as fatigue ductility coefficient, is denoted by ε'_f. The cyclic strength coefficient K' can be expressed in terms of σ'_f and ε'_f

$$K' = \frac{\sigma'_f}{(\varepsilon'_f)^{n'}}$$

[4.19]

Thereby Equation (4.17) takes the form

$$\frac{\Delta \varepsilon_p}{2} = \varepsilon'_f \left(\frac{\Delta \sigma}{2\sigma'_f} \right)^{\frac{1}{n'}}$$

[4.20]

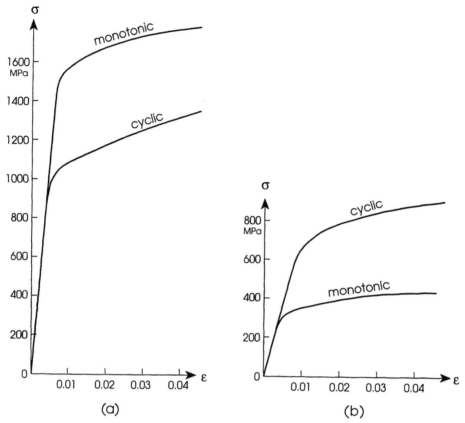

Figure 4.17 Cyclic and monotonic stress-strain curves for (a) steel SAE 4142, BHN 450, indicating cycle softening and (b) aluminum 2024-T4, indicating cycle hardening. (Data from "SAE Handbook," Reference 7.)

and Equation (4.18) becomes

$$\frac{\Delta\varepsilon}{2} = \frac{\Delta\sigma}{2E} + \varepsilon_f'\left(\frac{\Delta\sigma}{2\sigma_f'}\right)^{\frac{1}{n'}} \qquad\qquad [4.21]$$

This is the final mathematical expression of the cyclic stress-strain correlation derived by Morrow. (See Reference 8.)

For practical application regarding cyclic properties, one can use the following approximations for steels.[10] The fatigue strength coefficient for steels with BHN < 500 can be approximated by

$$\sigma_f' = S_u + 50,000\,\text{psi} \qquad\qquad [4.22]$$

The fatigue ductility coefficient can be approximated by

$$\varepsilon_f' = \varepsilon_f = \ln\frac{100}{100 - \%\,\text{RA}} \qquad\qquad [4.23]$$

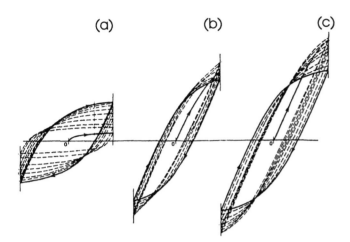

Figure 4.18 Cyclic behavior of OFHC coppers for
(a) fully annealed copper after 8060 reversals, for
(b) partially annealed copper after 4400 reversals
and for (c) cold-worked copper after 2000 reversals
(Data from Morrow, J., Reference 8. Reprinted
with permission.)

where %RA is the percentage of reduction in cross-sectional area at fracture, defined as

$$\% RA = 100 \frac{A_0 - A_f}{A_0}$$ [4.24]

For most metals the cyclic strain hardening exponent n' is

$$0.10 \leq n' \leq 20$$ [4.25]

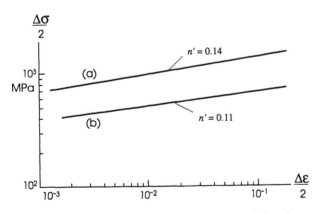

Figure 4.19 Cyclic stress-strain curves in log-log
coordinates for (a) steel SAE 4340 and
(b) aluminum 2024-T4. (Data from Endo, T.,
and J. Morrow, Reference 9.)

For the reader's convenience, cyclic properties of selected metals are given in the Appendix.

The mathematical derivation of the cyclic stress strain correlation can be schematically illustrated in Figure 4.20 using log-log coordinates. In the figure, Equations (4.16) and (4.20) are presented by straight lines AB and CD, respectively. The stress-strain correlation, Equation (4.21), is represented by curve AEFD. The curve exhibits three different behaviors: section AE reflects domination of elastic strain with plastic strain negligibly small; section FD, to the contrary, demonstrates domination of plastic strain with insignificant elastic strain; the intermediate section EF has both components equally vital.

To emphasize the fact that in practical design application one cannot substitute the monotonic correlation to solve problems of cyclic loading, let us consider the following case concerning fatigue of a machine part. Given a part, made of SAE 4142 steel, heat treated to 450 BHN, subject to fluctuating stresses with amplitude of 1200 MPa, Figure 4.21 shows the monotonic and cyclic stress-strain curves. Using the monotonic curve for design decision may lead to incorrectly determined safe limits. As can be seen from the figure, the stress amplitude is well below the yield point with the strain of about 0.007. Actually, the yield point of the material is much lower as a consequence of cyclic loading: the stress amplitude is above the yield point and the strain is about 0.020. If the cyclic data are disregarded the faulty design may lead to an unexpected failure of the part.

4.1.3 Fatigue life

The method of fatigue life prediction from the magnitude of the fluctuating strain has its origin in the stress method. It began with Basquin[11] who introduced an empirical relation based on the stress method (the law of endurance test)

$$S = BN^{-a} \hspace{4cm} [4.26]$$

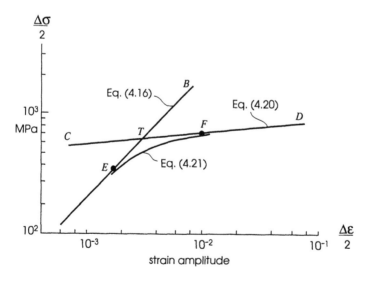

Figure 4.20 Cyclic stress-strain curve of RQC-100 steel in log-log coordinates. (Data from Tucker, L. and S. Bussa, see Reference 6.)

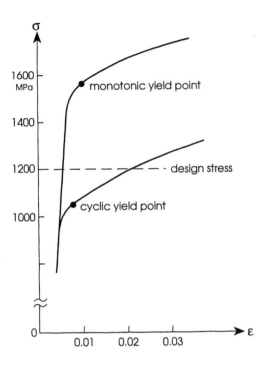

Figure 4.21 Fatigue design case for SAE 4142 steel heat treated to 450 BHN.

where S is the maximum stress and a and B are material properties. The relation was later revised and used to derive a strain-fatigue life correlation as follows.

Equation (4.26) can be modified to include true stress amplitude $\Delta\sigma/2$ in lieu of stress S and become

$$\frac{\Delta\sigma}{2} = \sigma'_f\left(2N_f\right) \qquad\qquad [4.27]$$

N_f is the number of load fluctuations up to the instance of specimen failure and $2N_f$ denotes the number of load reversals. The equation was confirmed by experimental data[12] throughout the fatigue range $N_f = 10$ to 10^6, as shown in Figure 4.22.

Equation (4.27) permits us to redefine the elastic and plastic strains, Equations (4.16) and (4.20), respectively, in terms of the number of reversals, $2N_f$. The elastic strain now becomes

$$\frac{\Delta\varepsilon_e}{2} = \frac{\Delta\sigma}{2E} = \frac{\sigma'_f}{E}\left(2N_f\right)^b \qquad\qquad [4.28]$$

while the plastic strain is

$$\frac{\Delta\varepsilon_p}{2} = \varepsilon'_f\left(\frac{\Delta\sigma}{2\sigma'_f}\right)^{\frac{1}{n'}} = \varepsilon'_f\left(2N_f\right)^c \qquad\qquad [4.29]$$

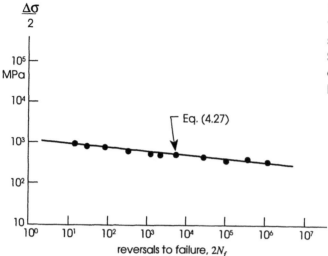

Figure 4.22 Stress amplitude versus cyclic life for annealed steel SAE 4340 (Data from Smith, R. W., M. H. Hirschberg, and S. S. Manson, see Reference 12.)

b is the fatigue strength coefficient and c the fatigue ductility exponent defined by

$$c = \frac{b}{n'}$$ [4.30]

Equation (4.29), called the Manson-Coffin rule, was proven by tests as shown in Figure 4.23. (See Reference 12.)

To arrive at the final correlation between the fatigue life and the total strain, we join Equations (4.28) and (4.29), as follows:

$$\frac{\Delta \varepsilon}{2} = \frac{\sigma_f'}{E}\left(2N_f\right)^b + \varepsilon_f'\left(2N_f\right)^c$$ [4.31]

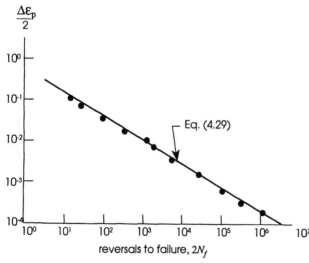

Figure 4.23
Plastic strain amplitude versus cyclic life for annealed steel SAE 4340 (Data from Smith, R. W., M. H. Hirschberg, and S. S. Manson, see Reference 12.)

The equation forms the basis of the strain method and was confirmed by many experiments. Figure 4.24 shows typical test data fitting the above equation. (See Reference 12.)

Figure 4.25 presents a schematic interpretation of Equation (4.31) using log-log coordinates. In the figure the above final correlation is a curve formed by joined elastic and plastic lines as per Equations (4.28), (4.29) and (4.31). Cross point T, between the elastic and plastic lines, divides the total strain range. This cross point occurs at what is called transition life $2N_t$. and it marks a transition from a plastic domain into an elastic one. To the left of transition life $2N_t$. the deformation is mainly plastic with only minor elastic strain. To the right, the deformation is mainly elastic with insignificant plastic strain. The lesser components on both sides increase as the number of reciprocals to failure, $2N_f$, gets closer to transition life $2N_t$. At this point the elastic and plastic strains become equal

$$\frac{\Delta \varepsilon_e}{2} = \frac{\Delta \varepsilon_p}{2} \qquad [4.32]$$

or

$$\frac{\sigma_f'}{E}\left(2N_f\right)^b = \varepsilon_f'\left(2N_f\right)^c \qquad [4.33]$$

It follows that the transition life is

$$2N_t = \varepsilon_f'\left(\frac{E\varepsilon_f'}{\sigma_f'}\right)^{\frac{1}{b-c}} \qquad [4.34]$$

In practice, integers denoting transition lives of different metals are useful in cases where an approximate range of the life of a machine part is indicated. If an estimated fatigue

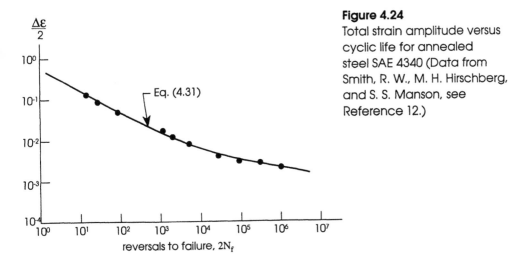

Figure 4.24

Total strain amplitude versus cyclic life for annealed steel SAE 4340 (Data from Smith, R. W., M. H. Hirschberg, and S. S. Manson, see Reference 12.)

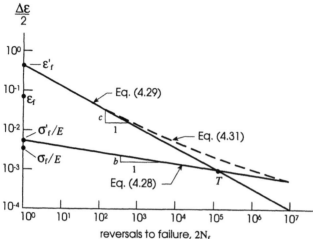

Figure 4.25 Plastic and elastic strain-life curves where σ_f is the monotonic fracture stress, while σ'_f is the cyclic fracture stress.

life of a part is smaller than the corresponding integer, Equation (4.31) can be replaced by a simpler Equation (4.29). For longer lives, Equation (4.28) substitutes for Equation (4.31). Figure 4.26 shows the transition lives of steels, as function of the Brinell Hardness Number (BHN). (See Reference 12.)

Stress concentration

Throughout this chapter, we have discussed the fatigue analysis of a smooth specimen. We now consider a specimen with a notch which acts as a stress raiser and presents a three-dimensional disturbance. This fact requires an extension of our analysis to include the determination of the total strain in the critical location within the notch. Even though we are dealing with a three-dimensional problem, let us for now disregard the true distribution and use a stress concentration factor as if this is a one-dimensional problem. (See Chapter 3.)

When all stresses are within elastic range, one can state

$$\sigma = K_t S \qquad\qquad [4.35]$$

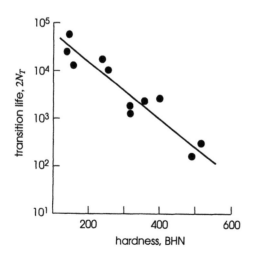

Figure 4.26 Transition fatigue life for steel as a function of hardness (Data from Smith, R. W., M. H. Hirschberg, and S. S. Manson, Reference 12.)

where K_t is the theoretical stress concentration factor, S is the nominal stress and σ the peak stress. An equivalent equation can be written for the elastic strains,

$$\varepsilon = K_t e \qquad [4.35']$$

When peak stress σ is higher than the yield point, we encounter a local plastic deformation resulting in a nonlinear form of the stress-strain relation. Here the local true stress σ and true strain ε have different concentration factors. Therefore, instead of Equations (4.35) and (4.35') we now have

$$\sigma = K_\sigma S \qquad [4.36]$$

and

$$\varepsilon = K_\varepsilon e \qquad [4.37]$$

assuming the nominal stress S is elastic. Concentration factors K_σ and K_ε are interdependent which can be expressed by means of numerous empirical relations.[13] One of them, the most prevalent, is the Neuber rule[14]

$$K_\sigma K_\varepsilon = K_t^2 \qquad [4.38]$$

Equation (4.38) was originally related to a specific problem: a curved notch subjected to shear loading. However, even though its general validity has not yet been proven theoretically, experimental evidence[15,16] shows its applicability in other cases.

In application to low-cycle fatigue, using Equations (4.36) and (4.37), Neuber's rule can be expressed as

$$\frac{\Delta\sigma}{\Delta S}\frac{\Delta\varepsilon}{\Delta e} = K_t^2 \qquad [4.39]$$

which is equivalent to

$$\frac{\Delta\sigma}{2}\frac{\Delta\varepsilon}{2} = \left(K_t\frac{\Delta S}{2}\right)^2\frac{1}{E} \qquad [4.40]$$

The above presents a hyperbolic relationship of $\Delta\varepsilon/2$ and $\Delta\sigma/2$. Since $\Delta\varepsilon/2$ and $\Delta\sigma/2$ are also connected by the equation

$$\frac{\Delta\varepsilon}{2} = \frac{\Delta\sigma}{2E} + \varepsilon'_f\left(\frac{\Delta\sigma}{2\sigma'_f}\right)^{\frac{1}{n'}} \qquad [4.41]$$

The total strain in the notched part, $\Delta\varepsilon/2$, can be determined solving the two above equations. Figure 4.27 presents a graphical solution where the hyperbola is based on

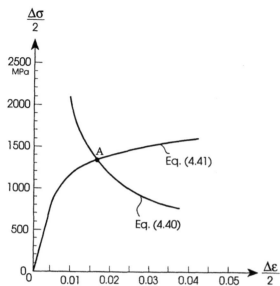

Figure 4.27 Determination of strain at the root of a notch according to Neuber rule.

Equation (4.40) and the stress-strain curve is based on Equation (4.41). Cross point A gives the total strain needed to compute the fatigue life.

Now the above illustrated methods to determine the fatigue life for a specimen without stress concentration can be applied to a notched specimen, including Figure 4.25 and Equations (4.28), (4.29) and (4.31).

Mean stress

For a more extensive analysis of fatigue life under cyclic loading, consider a specimen that is subjected to a completely reversed load with a superimposed steady stress. The resulting stress fluctuates between the maximum and minimum stresses. See Figure 4.28. The fatigue analysis can be performed by one of the three procedures presented below.

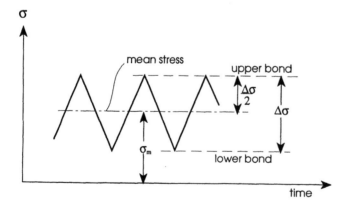

Figure 4.28 Fluctuating stress between an upper and lower bonds with a mean equal to the steady stress.

Manson's procedure   This procedure[17] introduces an equivalent stress amplitude, σ_N, with zero mean stress. The mathematical derivation is as follows. Let $\Delta\sigma/2$ be a given stress amplitude and σ_m the mean stress. The equivalent stress amplitude equals

$$\sigma_N = \sigma'_f \frac{\Delta\sigma/2}{\sigma'_f - \sigma_m} \tag{4.42}$$

See Equation (3.63). A cyclic process with zero mean, where the amplitude equals σ_N, will have the same life N_f as a process with the pair of stresses $(\Delta\sigma/2, \sigma_m)$. Figure 4.29 shows lines of constant life in coordinates σ_m and $\Delta\sigma/2$ which are assumed to be straight. Equivalent stress amplitude σ_N lies on the same constant-life line as point A $(\Delta\sigma/2, \sigma_m)$. Stress amplitude $\Delta\sigma/2$ equals

$$\frac{\Delta\sigma}{2} = \sigma_N \frac{\sigma'_f - \sigma_m}{\sigma'_f} \tag{4.43}$$

The correlation between equivalent stress amplitude σ_N and fatigue life, as follows from Equation (4.27), is

$$\sigma_N = \sigma'_f (2N_f)^b \tag{4.44}$$

Using Equations (4.43) and (4.44) we get the relationship between the elastic strain amplitude $\Delta\varepsilon_e/2$ and the fatigue life

$$\frac{\Delta\varepsilon_e}{2} = \frac{\Delta\sigma}{2E} = \frac{\sigma'_f - \sigma_m}{\sigma'_f} (2N_f)^b \tag{4.45}$$

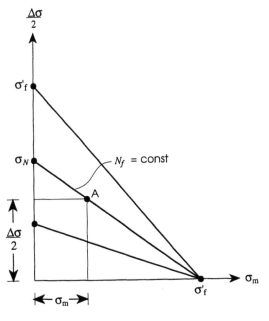

Figure 4.29 Lines of constant life in $(\Delta\sigma/2, (\sigma_m)$ coordinates.

To obtain the relationship between the plastic strain amplitude $\Delta\varepsilon_p/2$ and the fatigue life, we use Equation (4.20) in conjunction with Equations (4.43) and (4.44):

$$\frac{\Delta\varepsilon_p}{2} = \varepsilon_f' \left(\frac{\Delta\sigma}{2\sigma_f'}\right)^{\frac{1}{n'}} = \varepsilon_f' \left(\frac{\sigma_N}{\sigma_f'}\frac{\sigma_f'-\sigma_m}{\sigma_f'}\right)^{\frac{1}{n'}} \qquad [4.46]$$

$$= \varepsilon_f' \left[\left(2N_f\right)^b \frac{\sigma_f'-\sigma_m}{\sigma_f'}\right]^{\frac{1}{n'}}$$

$$= \varepsilon_f' \left(\frac{\sigma_f'-\sigma_m}{\sigma_f'}\right)^{\frac{1}{n'}}\left(2N_f\right)$$

By adding Equations (4.45) and (4.46), we get the final correlation between the strain amplitude $\Delta\varepsilon/2$ and the fatigue life as derived by Manson,

$$\frac{\Delta\varepsilon}{2} = \frac{\sigma_f'-\sigma_m}{E}\left(2N_f\right)^b + \varepsilon_f'\left(\frac{\sigma_f'-\sigma_m}{\sigma_f'}\right)^{\frac{1}{n'}}\left(2N_f\right)^c \qquad [4.47]$$

Figure 4.30 displays the above relationship in graphic form. Note that the Manson's procedure does not take into account stress relaxation at high plastic strains. (See discussion on relaxation at high plastic strains below.)

Morrow procedure   This procedure[18] is based upon a hypothesis that maximum stress σ_{max} equals the fatigue strength coefficient σ'_f when the fatigue life $N_f = 1$. Accordingly, the largest possible amplitude cannot be more than $(\sigma'_f-\sigma_m)$. It follows that at other values of fatigue life

$$2N_f = \left(\frac{\Delta\sigma/2}{\sigma_f'-\sigma_m}\right)^{\frac{1}{b}} \qquad [4.48]$$

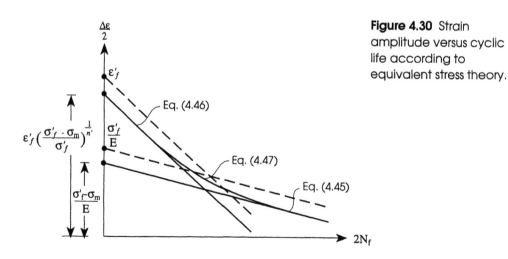

Figure 4.30 Strain amplitude versus cyclic life according to equivalent stress theory.

Consequently the stress versus fatigue-life correlation becomes

$$\frac{\Delta\sigma}{2} = (\sigma'_f - \sigma_0)\left(2N_f\right)^b \qquad\qquad [4.49]$$

and the elastic strain amplitude equals

$$\frac{\Delta\varepsilon_e}{2} = \frac{(\sigma'_f - \sigma_m)}{E}\left(2N_f\right)^b \qquad\qquad [4.50]$$

Similarly, when the fatigue life $N_f = 1$, the plastic strain amplitude $\Delta\varepsilon_p/2$ will reach ε'_f. Therefore

$$\frac{\Delta\varepsilon_p}{2} = \varepsilon'_f\left(\frac{\Delta\sigma/2}{\sigma'_f - \sigma_m}\right)^{\frac{1}{n'}} \qquad\qquad [4.51]$$

It follows from Equations (4.49) and (4.51) that the correlation between the plastic strain amplitude $\Delta\varepsilon_p/2$ and the fatigue life is

$$\frac{\Delta\varepsilon_p}{2} = \varepsilon'_f\left[\frac{(\sigma'_f - \sigma_m)\left(2N_f\right)^b}{\sigma'_f - \sigma_m}\right]^{\frac{1}{n'}} = \varepsilon'_f\left(2N_f\right)^c \qquad\qquad [4.52]$$

By adding Equations (4.50) and (4.52) we obtain the fatigue life correlation

$$\frac{\Delta\varepsilon}{2} = \frac{(\sigma'_f - \sigma_m)}{E}\left(2N_f\right)^b + \varepsilon'_f\left(2N_f\right)^c \qquad\qquad [4.53]$$

Figure 4.31 shows the above relationship in graphic form. Note that according to Morrow's theory the effect of the mean stress is restricted to the elastic strain domain.

   Berkovits' procedure   This method[19] adapts the Goodman theory of low-cycle fatigue. The basic correlation of Goodman is

$$\frac{S_a}{S_f} + \frac{S_m}{S_u} = 1 \qquad\qquad [4.54]$$

See Equation (3.41). When presenting the Goodman equation in terms of true stresses, one expects it to become

$$\frac{\Delta\sigma/2}{\left(\Delta\sigma/2\right)_f} + \frac{\sigma_m}{\sigma'_f} = 1 \qquad\qquad [4.55]$$

where $(\Delta\sigma/2)_f$ denotes the fatigue limit. Berkovits established the fact that the apex of

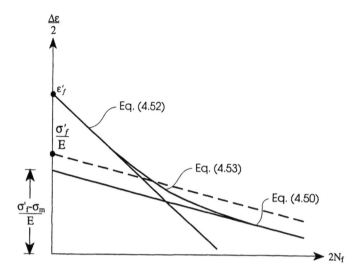

Figure 4.31 Strain amplitude versus cyclic life according to maximum stress theory.

Goodman line, σ'_m, should be located before σ'_f. See Figure 4.32. Thus the corrected correlation becomes

$$\frac{\Delta\sigma/2}{\left(\Delta\sigma/2\right)_f} + \frac{\sigma_m}{\sigma'_m} = 1 \qquad [4.56]$$

By introducing a mean stress sensitivity coefficient

$$k = \frac{\sigma'_f}{\sigma'_m} \qquad [4.57]$$

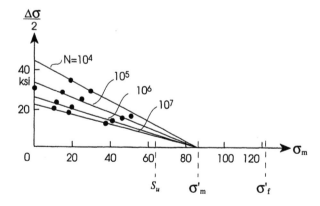

Figure 4.32 Construction of appex (σ_m for Incoloy 901 in Goodman diagram in true stress and strain coordinates. (Data from Fang, D., and A. Berkovits, see Reference 19.)

Equation (4.56) becomes

$$\frac{\Delta\sigma/2}{\left(\Delta\sigma/2\right)_f} + \frac{k\sigma_m}{\sigma_f'} = 1 \qquad [4.58]$$

It follows that the original mean stress σ_m had to be multiplied by mean stress sensitivity factor k to become $k\sigma_m$.

Returning to the Manson procedure above, Equation (4.42), defining equivalent stress amplitude σ_N, has also to include corrected term $k\sigma_m$. The equivalent stress amplitude now becomes

$$\sigma_N = \sigma_f' \frac{\Delta\sigma/2}{\sigma_f' - k\sigma_m} \qquad [4.59]$$

Following the same steps as before, we finally get

$$\frac{\Delta\varepsilon_e}{2} = \frac{\sigma_f' - k\sigma_m}{\sigma_f'}\left(2N_f\right)^b \qquad [4.60]$$

$$\frac{\Delta\varepsilon_p}{2} = \varepsilon_f'\left(\frac{\sigma_f' - k\sigma_m}{\sigma_f'}\right)^{\frac{1}{n'}}\left(2N_f\right)^c \qquad [4.61]$$

and

$$\frac{\Delta\varepsilon}{2} = \frac{\sigma_f' - k\sigma_m}{E}\left(2N_f\right)^b + \varepsilon_f'\left(\frac{\sigma_f' - k\sigma_m}{\sigma_f'}\right)^{\frac{1}{n'}}\left(2N_f\right)^c \qquad [4.62]$$

The equation was confirmed by tests. (See Reference 19.)

Stress concentration It was suggested that the above described procedures could be used in conjunction with the Neuber rule, to include parts subjected simultaneously to mean stress and stress concentration. Some experiments seem to confirm the applicability but, since the data examined to date are too specific, a wider application should be made with caution. (See Reference 16.)

Cyclic creep and stress relaxation In tests at very high strain amplitudes, two different phenomena may be observed: cyclic creep and stress relaxation. Cyclic creep is a progressive straining (ratcheting) under stress control with a tensile mean stress, as shown in Figure 4.33. The figure reflects a case where the stress amplitude and the mean stress are kept constant, while the mean strain increases continuously due to creep. As seen from the figure, the unstable hysteresis moves continuously forward in the direction of increasing strain. In a cycle-softening metal the strain continuously increases until the final failure. In a cycle-hardening metal the strain decreases (see Reference 8), the

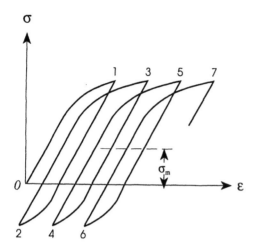

Figure 4.33 Cyclic creep under stress control with tensile mean stress.

cyclic creep decelerates and may cease after a number of cycles. Stress relaxation may occur when a specimen is subjected to a cyclic load with strain control and tensile mean strain. See Figure 4.34. The large plastic strain amplitude causes the mean stress to relax. The higher the strain, the faster the relaxation rate until it reaches zero.

Both phenomena are known to occur at strain amplitude, $\Delta\varepsilon/2$, larger than $0.005 - 0.01$, with fatigue lives of less than 100. Their appearance depends on metal properties and load control.

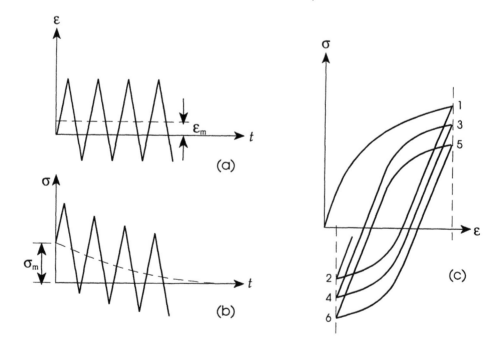

Figure 4.34 Stress relaxation under strain control: (a) strain versus time; (b) stress versus time; (c) stress versus strain.

4.2
MULTIAXIAL ANALYSIS

The fatigue analysis of multiaxial fluctuating loading involves three stages: first, derivation of fluctuating multiaxial stress and strain components; second, replacement of these components by one-dimensional equivalents; and third, computation of fatigue life utilizing the one-dimensional methods described in Section 4.1.

The three stages described below are based on continuous assumption used in Chapter 3 that the stress and strain components fluctuate in phase, originating from the same source.

4.2.1 Multiaxial stress and strain components

By multiaxial we mean two- and three-dimensional components. However, in our discussion we refer only to three-dimensional space, due to the fact that one can adapt the results to two dimensions.

There are two approaches to obtain the fluctuating three-dimensional stress and strain components as function of time. The history of fluctuating loading comprises loading and unloading stages. The approach to compute the stress and strain components depends on conditions that exist during a loading stage as plastic deformation sets in. At this stage stress components rise either in the same ratio—proportional loading (the loading which exists, for example, in a symmetric pressure vessel under increasing pressure); or in nonequal ratios—nonproportional loading, (the loading which may take place in a flat plate under plane strain conditions). See Section 2.2.4. Nonproportional loading requires an incremental analysis, while proportional loading may be treated simply by using the deformation theory of plasticity, as illustrated in detailed examples computing the fatigue life, in Section 4.2.3, below.

Nonproportional loading Figure 4.35 illustrates nonproportional loading. It shows a flat plate with a notch, where stresses in *z*-direction rise not in proportion to

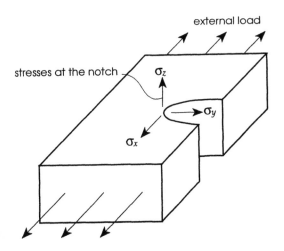

Figure 4.35 Stresses at the notch in a plane strain problem in nonproportional loading.

the stresses in *xy*-plane. For the purpose of computation of nonproportional loading, the load application under this condition is viewed as a succession of small discrete steps. At each step the stress and strain increments, $\delta\sigma_{ij}$ and $\delta\varepsilon_{ij}$, are computed according to the elasticity and plasticity theories, Equations (2.32) and (2.71), respectively. Based on this, the strain increments equal

$$\delta\varepsilon_{ij} = \delta\varepsilon_{ij}^e + \delta\varepsilon_{ij}^p \qquad\qquad\qquad [4.63]$$

The cumulative stress and strain components are obtained using the following equations

$$\sigma_{ij} = \sum \delta\sigma_{ij}, \quad \varepsilon_{ij} = \sum\left(\delta\varepsilon_{ij}^e + \delta\varepsilon_{ij}^p\right) \qquad\qquad [4.64]$$

The amplitudes of the alternating three-dimensional stress and strain components are computed from

$$\frac{\Delta\sigma_{ij}}{2} = \frac{\sigma_{ij,\max} - \sigma_{ij,\min}}{2}, \quad \frac{\Delta\varepsilon_{ij}}{2} = \frac{\varepsilon_{ij,\max} - \varepsilon_{ij,\min}}{2} \qquad [4.65]$$

while the mean values are computed from

$$\sigma_{ij,m} = \frac{\sigma_{ij,\max} + \sigma_{ij,\min}}{2}, \quad \varepsilon_{ij,m} = \frac{\varepsilon_{ij,\max} + \varepsilon_{ij,\min}}{2} \qquad [4.66]$$

To evaluate the cumulative stress and strain components the Cartesian components are used instead of the principal ones, as can be seen from the above equations. The necessity for this comes as a result of nonproportionality—the change in principal directions at each incremental step.

Proportional loading As mentioned earlier, the proportional loading may be treated simply by using the deformation theory of plasticity—the stress and strain components are increasing in a constant ratio (see Section 2.2). The components at the end of a loading stage are computed directly in integral form where each loading is considered as one large increment. The unloading is viewed as a separate stage, analyzed applying the elasticity theory.

4.2.2 Equivalent strains and stresses

The fundamental approaches to derive the equivalent stresses and strains were discussed in Chapter 2. Knowing the three-dimensional stress and strain components, one can extend the application of equivalent stresses and strains to fluctuating loading.[20-22] Presented below is the adaptation of known criteria to derivation of equivalent alternating strains and equivalent mean stresses—criteria of von Mises and of Tresca, discussed in Chapter 2, and that of Sines, introduced in Chapter 3. While the first one demands a more elaborate adaptation, the other two, Tresca and Sines, are simpler.

Equivalent alternating strains

Application of von Mises criterion The amplitudes of alternating three-dimensional strain components contain the elastic and plastic parts

$$\frac{\Delta\varepsilon_{ij}}{2} = \frac{\Delta\varepsilon_{ij}^e}{2} + \frac{\Delta\varepsilon_{ij}^p}{2}$$

[4.67]

Correspondingly, the equivalent strain amplitude also contain two parts

$$\frac{\Delta\varepsilon}{2} = \frac{\Delta\varepsilon_e}{2} + \frac{\Delta\varepsilon_p}{2}$$

[4.68]

Based on derivations presented in Sections 2.1 and 2.2., these parts equal

$$\frac{\Delta\varepsilon_e}{2} = \frac{1}{(1+\nu)\sqrt{2}} \left[\left(\frac{\Delta\varepsilon_1^e}{2} - \frac{\Delta\varepsilon_2^e}{2}\right)^2 + \left(\frac{\Delta\varepsilon_2^e}{2} - \frac{\Delta\varepsilon_3^e}{2}\right)^2 + \left(\frac{\Delta\varepsilon_3^e}{2} - \frac{\Delta\varepsilon_1^e}{2}\right)^2 \right]^{\frac{1}{2}}$$

[4.69]

and

$$\frac{\Delta\varepsilon_p}{2} = \frac{\sqrt{2}}{3} \left[\left(\frac{\Delta\varepsilon_1^p}{2} - \frac{\Delta\varepsilon_2^p}{2}\right)^2 + \left(\frac{\Delta\varepsilon_2^p}{2} - \frac{\Delta\varepsilon_3^p}{2}\right)^2 + \left(\frac{\Delta\varepsilon_3^p}{2} - \frac{\Delta\varepsilon_1^p}{2}\right)^2 \right]^{\frac{1}{2}}$$

[4.70]

where the principal strain amplitudes are derived from the component strain amplitudes, Equation (4.65).

Equation (4.70) follows from Equation (4.69), when one uses the Poisson ratio for plastic strain, $\nu = 1/2$. Correspondingly, using an average Poisson ratio, one gets the equivalent total strain amplitude

$$\frac{\Delta\varepsilon}{2} = \frac{1}{(1+\bar\nu)\sqrt{2}} \left[\left(\frac{\Delta\varepsilon_1}{2} - \frac{\Delta\varepsilon_2}{2}\right)^2 + \left(\frac{\Delta\varepsilon_2}{2} - \frac{\Delta\varepsilon_3}{2}\right)^2 + \left(\frac{\Delta\varepsilon_3}{2} - \frac{\Delta\varepsilon_1}{2}\right)^2 \right]^{\frac{1}{2}}$$

[4.71]

where

$$\bar\nu = \frac{\nu\frac{\Delta\varepsilon_e}{2} + 0.5\frac{\Delta\varepsilon_p}{2}}{\frac{\varepsilon_e}{2} + \frac{\varepsilon_p}{2}}$$

[4.72]

When applicable, Equation (4.71) can be stated using Cartesian strain components

$$\frac{\Delta\varepsilon}{2} = \frac{1}{(1+\bar{\nu})\sqrt{2}} \left\{ \left[\left(\frac{\Delta\varepsilon_x}{2} - \frac{\Delta\varepsilon_y}{2} \right)^2 + \left(\frac{\Delta\varepsilon_y}{2} - \frac{\Delta\varepsilon_z}{2} \right)^2 + \left(\frac{\Delta\varepsilon_z}{2} - \frac{\Delta\varepsilon_x}{2} \right)^2 \right] + 6\left[\left(\frac{\Delta\gamma_{xy}}{4} \right)^2 + \left(\frac{\Delta\gamma_{yz}}{4} \right)^2 + \left(\frac{\Delta\gamma_{zx}}{4} \right)^2 \right] \right\}^{\frac{1}{2}} \tag{4.73}$$

Application of Tresca criterion   Here the equivalent strain amplitude equals the largest of the three

$$\frac{\Delta\varepsilon}{2} = \frac{1}{1+\bar{\nu}} \left| \frac{\Delta\varepsilon_1}{2} - \frac{\Delta\varepsilon_2}{2} \right| \quad \text{or} \quad \frac{1}{1+\bar{\nu}} \left| \frac{\Delta\varepsilon_2}{2} - \frac{\Delta\varepsilon_3}{2} \right| \quad \text{or} \quad \frac{1}{1+\bar{\nu}} \left| \frac{\Delta\varepsilon_3}{2} - \frac{\Delta\varepsilon_1}{2} \right| \tag{4.74}$$

There are other known theories that are used occasionally as criteria for defining the equivalent alternating strain, as for instance the criterion of maximum principal strain (see Reference 20) expressed as

$$\frac{\Delta\varepsilon}{2} = \frac{\Delta\varepsilon_1}{2} \tag{4.75}$$

Equivalent mean stresses

Application of von Mises criterion   The equivalent mean stress is

$$\sigma_{m,e} = \sqrt{\frac{1}{2}\left[\left(\sigma_{1,m} - \sigma_{2,m}\right)^2 + \left(\sigma_{2,m} - \sigma_{3,m}\right)^2 + \left(\sigma_{3,m} - \sigma_{1,m}\right)^2 \right]} \tag{4.76}$$

where $\sigma_{1,m}$, $\sigma_{2,m}$ and $\sigma_{3,m}$ are the principal mean stress components.

Application of Tresca criterion   The equivalent mean stress is the largest of the three

$$\sigma_{m,e} = \left|\sigma_{1,m} - \sigma_{2,m}\right| \quad \text{or} \quad \left|\sigma_{2,m} - \sigma_{3,m}\right| \quad \text{or} \quad \left|\sigma_{3,m} - \sigma_{1,m}\right| \tag{4.77}$$

Application of Sines criterion   Even though this is applicable to high cycle fatigue analysis, we shall adapt it to the low cycle, as used in practice. Let us consider the simplified version of Sines criterion presented by Fuchs and Stephens.[23] Here the equivalent mean stress equals

$$\sigma_{m,e} = \sigma_{1,m} + \sigma_{2,m} + \sigma_{3,m} \tag{4.78}$$

The method of Sines appears to be more practical since it recognizes the fact that a mean tension stress shortens fatigue life, whereas a mean compressive stress extends it. The methods of von Mises and Tresca disregard the different influences of tension and compression stresses on the fatigue life. Note that in the above analysis the relaxation effects are disregarded, meaning that constant mean stresses are applied throughout the fatigue life.

4.2.3 Computing fatigue life

To illustrate the determination of fatigue life for two types of loading, proportional and nonproportional, two detailed examples follow. First, a bar in axial loading to reflect the proportional loading and, second, a perforated plate to illustrates the nonproportional loading. Application of the different criteria, given above, applied to the same conditions produces significant differences in both examples. This is to show that no general conclusion can be reached about the results in the use of any of the criteria. Design problems for fatigue life always require more than one solution. The judgment to choose the safest one remains the realm of the design engineer.

Bar subject to axial loading

Consider a bar, subject to fluctuating axial loading, tension and compression, with zero mean stress. See Figure 4.36. During the loading stage (considered separately from the unloading stage) the stress path is constant and the loading is proportional. The one existing fluctuating stress component is σ_1, so that the stress amplitude is

$$\frac{\Delta\sigma}{2} = \sigma_1 \tag{4.79}$$

Let us determine the fatigue life of the given bar applying each one of the adapted criteria of von Mises and Tresca. It is interesting to compare the obtained fatigue-life correlations, presented graphically in Figure 4.37.

Application of von Mises criterion The three-dimensional strain amplitudes comprise elastic and plastic parts

$$\frac{\Delta\varepsilon_i}{2} = \frac{\Delta\varepsilon_i^e}{2} + \frac{\Delta\varepsilon_i^p}{2} \quad \text{where} \quad i = 1, 2, 3 \tag{4.80}$$

The elastic strain components equal

$$\frac{\Delta\varepsilon_1^e}{2} = \frac{\Delta\sigma}{2E}, \quad \frac{\Delta\varepsilon_2^e}{2} = \frac{\Delta\varepsilon_3^e}{2} = -v\frac{\Delta\sigma}{2E} \tag{4.81}$$

The plastic strain components are

$$\frac{\Delta\varepsilon_i^p}{2} = \frac{3}{2}\frac{s_i}{\sigma_e}\frac{\Delta\varepsilon_p}{2} \tag{4.82}$$

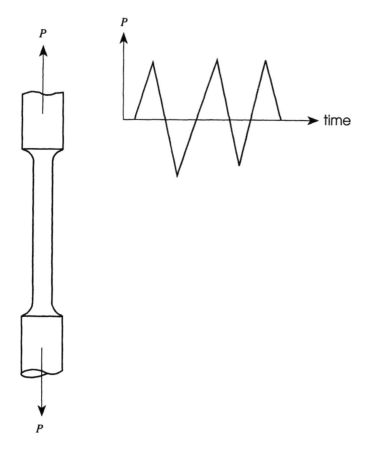

Figure 4.36 Bar under fluctuating axial loading.

With deviatoric stresses

$$s_1 = \frac{2}{3}\sigma_1 \quad , \quad s_2 = s_3 = -\frac{1}{3}\sigma_1 \qquad [4.83]$$

the plastic strain components become

$$\frac{\Delta\varepsilon_1^p}{2} = \frac{\Delta\varepsilon_p}{2}, \quad \frac{\Delta\varepsilon_2^p}{2} = \frac{\Delta\varepsilon_3^p}{2} = -0.5\frac{\Delta\varepsilon_p}{2} \qquad [4.84]$$

Consequently, using Equations (4.69) and (4.70), we obtain the equivalent strain amplitude

$$\frac{\Delta\varepsilon}{2} = \frac{\Delta\sigma}{2E} + \frac{\Delta\varepsilon_p}{2} \qquad [4.85]$$

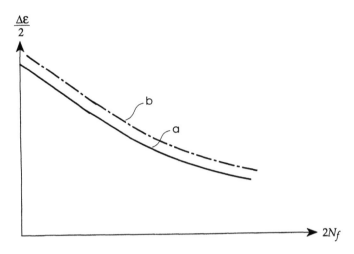

$\dfrac{\Delta\varepsilon}{2}$

$2N_f$

Figure 4.37 Strain amplitude versus cyclic life in axial loading using different criteria: (a) von Mises criterion; (b) Tresca criterion.

According to the one-dimensional theory, the elastic and plastic parts of Equation (4.85), respectively, equal

$$\frac{\Delta\sigma}{2E} = \frac{\sigma'_f}{E}\left(2N_f\right)^b \qquad [4.86]$$

and

$$\frac{\Delta\varepsilon_p}{2} = \varepsilon'_f\left(2N_f\right)^c \qquad [4.87]$$

Hence the fatigue life can be expressed, as in Equation (4.31)

$$\frac{\Delta\varepsilon}{2} = \frac{\sigma'_f}{E}\left(2N_f\right)^b + \varepsilon'_f\left(2N_f\right)^c \qquad [4.88]$$

Application of Tresca criterion   Applying the Tresca criterion, the equivalent strain amplitude equals

$$\frac{\Delta\varepsilon}{2} = \frac{1}{2}\left|\Delta\varepsilon_1 - \Delta\varepsilon_2\right| = \left(1+v\right)\frac{\Delta\sigma}{2E} + \frac{3}{2}\frac{\Delta\varepsilon_p}{2} \qquad [4.89]$$

The fatigue life, using Equations (4.86) and (4.87), differs here from Equation (4.88) and is expressed by a following correlation

$$\frac{\Delta\varepsilon}{2} = \left(1+v\right)\frac{\sigma'_f}{E}\left(2N_f\right)^b + \frac{3}{2}\varepsilon'_f\left(2N_f\right)^c \qquad [4.90]$$

The curves in Figure 4.37 show plainly the divergence in the solutions obtained from applying the two criteria in computation of fatigue life. In this example, the solution obtained with the use of von Mises criterion indicates shorter fatigue lives than obtained with the Tresca application.

Perforated plate

Consider a symmetric plate with a hole, made of steel SAE 5160, subject to cyclic loading as specified in Figure 4.38. Before deriving the fatigue life, let us compute the multiaxial stress and strain components, using an incremental procedure as per Equations (4.64). The solution is obtained by ANSYS program, using the finite element method in conjunction with Newton-Raphson iteration procedure for elasto-plastic materials. The computer generated plots are shown in Figures 4.39 to 4.41. Figure 4.39 shows elastic and plastic domains within the plate, at the end of a loading stage; MX indicates the point of maximum stress. Figures 4.40 and 4.41 show the histories of stress and strain components at point MX during loading and unloading. One can distinguish the changes in stress components in nonequal ratios. (The nonproportional changes may be explained by the fact that during the process of loading a change occurs in the con-

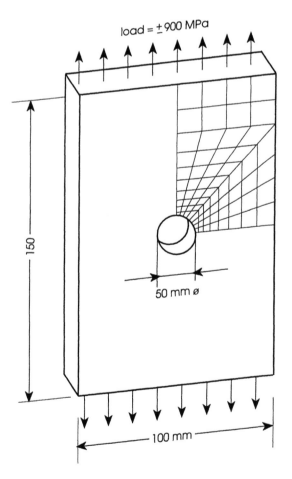

load = ± 900 MPa

150

50 mm ø

100 mm

Figure 4.38 Perforated plate under fluctuating load.

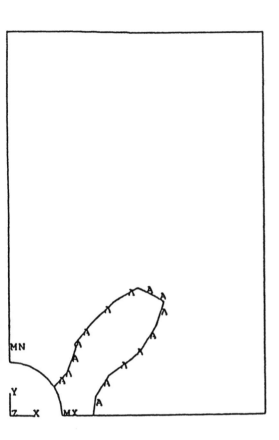

Figure 4.39 Plastic and elastic domains in perforated plate; the plastic domain is in the lower left corner.

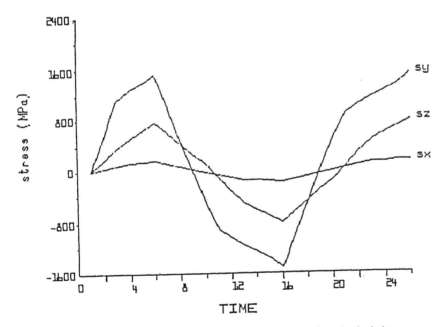

Figure 4.40 Stress history at critical location in the perforated plate.

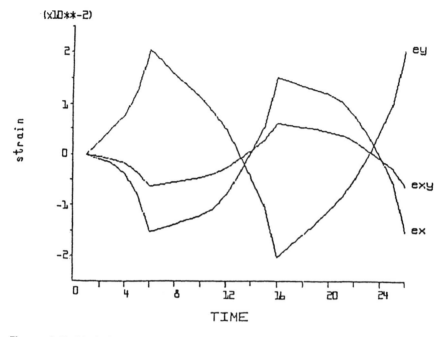

Figure 4.41 Strain history at critical location in the perforated plate.

ditions, from elastic to plastic, whereby the Poisson ratio increases from $v = 0.3$ to $v = 0.5$.) The numerically computed strain amplitudes are as follows:

$$\Delta\varepsilon_x/2 = 0.0153, \ \Delta\varepsilon_y/2 = 0.0204, \ \Delta\gamma_{xy}/2 = 0.00625$$

Now we proceed to determine the fatigue life of the plate. With the help of von Mises criterion, we can express the equivalent total strain amplitude as

$$\frac{\Delta\varepsilon}{2} = \frac{1}{(1+\bar{v})\sqrt{2}}\left\{(0.0153-0.0204)^2 +(0.0204-0)^2 +(0-0.0153)^2 + 6\left(\frac{0.00625}{2}\right)^2\right\}^{\frac{1}{2}}. \qquad [a]$$

To compute the total strain amplitude one has to know the average Poisson ratio. If we assume a Poisson ratio of 0.5, the strain amplitude becomes

$$\Delta\varepsilon/2 = 0.0124.$$

To arrive at the equation which will give us the fatigue life for the given plate, we insert the properties of SAE 5160 steel (see Appendix) into Equation (4.31)

$$0.0124 = \frac{1931}{193,000}\left(2N_f\right)^{-0.071} + 0.40\left(2N_f\right)^{-0.57} \qquad [b]$$

Solving the above equation, we get a fatigue life $2N_f = 1200$ reversals or 600 cycles.

On the hand if one uses the Tresca criterion, one obtains the principal strain amplitudes

$$\Delta\varepsilon_1/2 = 0.0259, \quad \Delta\varepsilon_2/2 = 0.0098, \quad \Delta\varepsilon_3/2 = 0$$

The resulting equivalent strain amplitude equals

$$\frac{\Delta\varepsilon}{2} = \frac{1}{1+\bar{\nu}} \left| \frac{\Delta\varepsilon_3}{2} - \frac{\Delta\varepsilon_1}{2} \right| = 0.0172 \qquad [c]$$

whereby the fatigue life becomes $2N_f = 600$ reversals or 300 cycles. This differs from the result obtained with von Mises criterion by a factor of 2. In this example, the results obtained with Tresca criterion (contrary to the example of the bar subject to axial loading) produce the more conservative values in computing the fatigue life.

REFERENCES

1. *ASME Boiler and Pressure Vessel Code.* 1989. Section 8, Rules for Construction of Pressure Vessels. New York: ASME.
2. "Fatigue Design Handbook." 1988. Rice, R.C. ed., Warrendale, PA: Society of Automotive Engineers.
3. "Constant-Amplitude Low-Cycle Fatigue Testing." 1980. Annual Book of ASTM Standards, Part 10, Philadelphia, PA: ASTM E606–80.
4. Hollomon, J.H. 1945. "Tensile Deformation," *Trans. Amer. Inst. Mining and Metallurgical Engineers (Iron and Steel Div.)*, 162, 268–290.
5. Dieter, Jr., G.E. 1986. *Mechanical Metallurgy.* New York: McGraw Hill.
6. Tucker, L., and S. Bussa. 1977. "The SAE Cumulative Fatigue Damage Test Program," *Fatigue Under Complex Loading*, Wetzel, R.M., ed. Warrendale, PA: SAE.
7. "SAE Handbook." 1989. Warrendale, PA: Society of Automotive Engineers.
8. Morrow, J. 1965. "Cyclic Plastic Strain Energy and Fatigue of Metals," *Internal Friction, Damping and Cyclic Plasticity*, ASTM STP 378, 45–87.
9. Endo, T., and J. Morrow. 1969. "Cyclic Stress-Strain and Fatigue Behavior of Representative Aircraft Metals, *J. Materials*, 4, 1, 159–175.
10. Socie, D.F., and J. Morrow. 1976. "Review of Contemporary Approaches to Fatigue Damage Analysis," FCP Report 24, College of Engineering, Univ. of Illinois.
11. Basquin, O.H. 1910. "The Exponential Law of Endurance Tests," *Proc. Amer. Soc. Test. Mater.,* 10, 625–630.
12. Smith, R.W., M.H. Hirschberg, and S.S. Manson. 1963. "Fatigue Behavior of Materials under Strain Cycling in Low and Intermediate Life Range," NASA Tech. Note, D-1574.
13. Klesnil, M., and P. Lukas. 1980. *Fatigue of Metallic Materials.* The Netherlands: Elsevier.
14. Neuber, H. 1961. "Theory of Stress Concentration for Shear-Strained Prismatic Bodies," *Trans. ASME J. Appl. Mech.*, 28, 544.
15. Peterson, R.E. 1963. "Fatigue of Metals, Part 3, Engineering and Design Aspects," *Materials Research and Standards*, 3, 122–139.
16. Wetzel, R.M. 1968. "Smooth Specimen Simulation of Fatigue Behavior of Notches," *J. Materials*, 3, 646–657.
17. Manson, S.S. 1960. "Effect of Mean Stress and Strain on Cyclic Life," *Machine Design*, (August 4) 129–135.
18. Morrow, J. 1968. "Fatigue Properties of Metals," Section 3.2 in *Fatigue Design Handbook*, J.A. Graham, ed., Warrendale, PA: SAE.
19. Fang, D., and A. Berkovits. 1993. "Mean Stress Models for Low-Cycle Fatigue of a Nickel-Base Superalloy *Int. J. Fatigue*, 16, 429–437.

20. Krempl, E. 1973. "The Influence of State of Stress on Low-Cycle Fatigue of Structural Materials," STP 549. Philadelphia: ASTM.
21. Garud, Y.S., 1981. "Multiaxial Fatigue: A Survey of the State of Art," *J. of Testing and Evaluation, 9*, 165–178.
22. *Multiaxial Fatigue.* 1985. Miller, K.J,. and M.W. Brown, ed., Philadelphia, PA: ASTM, STP 853.
23. Fuchs, H.O., and R.I. Stephens, 1980. *Metal Fatigue in Engineering*, New York: John Wiley.

5

CRACK PROPAGATION

The final stages of fatigue in a machine part under a fluctuating load are manifested by an onset, at a specific point in the part, of a crack. The initiation of a crack begins long before that with crystal imperfections and microscopic dislocations which are followed by an outright split, propagating into a visible crack that leads to an ultimate fracture. Chapters 3 and 4 discussed the stress and strain methods to estimate life expectancy. In this chapter is introduced the method of fracture mechanics which complements the strain method for more accurate results of an entire destruction process. Strain method concerns itself with the initial stage, while fracture mechanics analyzes the crack propagation in its entirety. The physical changes in the material follow from one stage to the other with many factors influencing the duration of one stage or the other. Theoretically we need to divide the timing of the process—from the start to a final fracture—into three discrete periods: (a) initial damage in a submicroscopic scale; (b) visible damage, cracks initiation and growth; (c) final instantaneous fracture which is the ultimate failure. It is apparent that the accuracy of specimen's life expectancy depends to a large degree upon prediction, as accurate as possible, of the indicated destruction period. Both the specimen's properties and the outer loading are factors that influence the analysis.

5.1
FRACTURE MECHANICS

5.1.1 Linear elastic fracture mechanics

Fracture mechanics is concerned with the process covering stages (b) visible damage and (c) final failure. An examination of the damaged surface of a broken part due to fatigue discloses stage (b) the fatigue zone and stage (c) rupture zone. See Figure 5.1. The fatigue zone, an area of the growing crack or cracks, is relatively smooth with the location of cracks' origins showing (in general) quite clearly. The surface in the rupture zone, on the other hand, is rough.

One of the known methods of analyzing this process is the method of linear elastic fracture mechanics (LEFM). It is designated to compute the crack propagation based on the fundamental assumption that the material is linearly elastic and the fracture is brittle. This method basically approaches the crack propagation problems in two-dimensional space. For more complex cases an adaptation of the basic correlations is used, as will be shown later.

Before attempting to analyze the crack propagation it is necessary to define the stresses and displacements in the area. Figure 5.2 shows an infinite plate with an existing crack, subject to tension. Shown in the figure is a coordinate system of the crack. The general form[1] of the Cartesian stress components in the vicinity of the crack is found to be

$$\sigma_{ij} = \frac{1}{\sqrt{r}}\left[c_0 f_{0ij}(\theta) + c_1 r f_{1ij}(\theta) + c_2 r^2 f_{2ij}(\theta) + \ldots\right] \qquad [5.1]$$

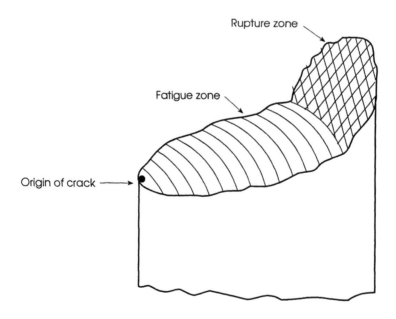

Figure 5.1 A typical fatigue fracture surface.

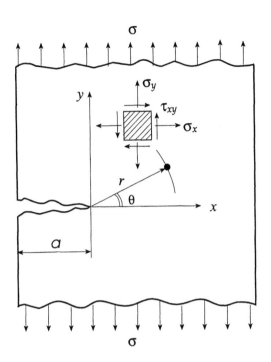

Figure 5.2 Coordinate system in the vicinity of a crack.

Attention is brought to the immediate vicinity of the crack tip where point (r, θ) is close to the tip, that is, $r < < a$. This fact allows us to disregard higher order terms using the first term only. Thus Equation (5.1) in the vicinity of the tip becomes

$$\sigma_{ij} = \frac{K}{\sqrt{2\pi r}} f_0(\theta) \qquad [5.2]$$

The displacement components, according to the same linear elastic solution, equal

$$u_i = \frac{1+v}{E} K \sqrt{\frac{2r}{\pi}} \phi_i(\theta) \qquad [5.3]$$

Here we introduced a constant K, called the stress intensity factor, which signifies the relationship of three factors: the geometry of the plate, the loading, and the length of the crack. The stress intensity factor is defined as follows

$$K = f\sigma\sqrt{\pi a} \qquad [5.4]$$

In the above, f is a compliance function that describes the geometry of the part and σ is a stress in a remote distance which designates the loading. See Figure 5.2. We will review here the behavior of the stress intensity factor in parts with different geometries.

Now let us proceed with the analysis of the stresses and displacements to include various loading modes. Consider Figure 5.3 with three different cases, defining each one as follows: Mode I, the opening mode, where the crack's sides move perpendicu-

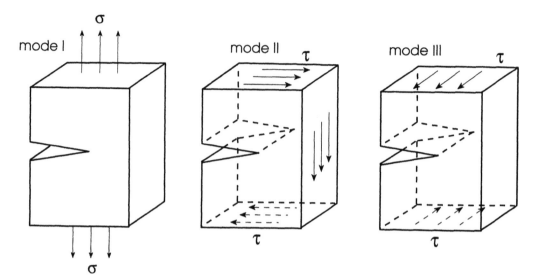

Figure 5.3 Three modes of loading in a crack.

larly apart; Mode II, the sliding mode, where the crack's sides slide in parallel within plane of the crack; and Mode III, the tearing mode, where the crack's sides are torn apart and the surfaces moved out of plane.

Each mode is approached individually. The stresses and the displacements for each mode are derived[2] correspondingly from Equations (5.2) and (5.3) as follows.

Mode I

Stresses in xy-plane:

$$\sigma_x = \frac{K_1}{\sqrt{2\pi r}} \cos\frac{\theta}{2}\left(1 - \sin\frac{\theta}{2}\sin\frac{3\theta}{2}\right), \qquad [5.5]$$

$$\sigma_y = \frac{K_1}{\sqrt{2\pi r}} \cos\frac{\theta}{2}\left(1 - \sin\frac{\theta}{2}\sin\frac{3\theta}{2}\right),$$

$$\tau_{xy} = \frac{K_1}{\sqrt{2\pi r}} \sin\frac{\theta}{2}\cos\frac{\theta}{2}\cos\frac{3\theta}{2};$$

Stresses in z-direction applicable to plane strain problems:

$$\sigma_z = \nu\left(\sigma_x + \sigma_y\right), \qquad [5.6]$$
$$\tau_{xz} = \tau_{yz} = 0;$$

Stresses in z-direction applicable to plane stress problems:

$$\sigma_z = \tau_{xz} = \tau_{yz} = 0 \qquad [5.7]$$

Displacements in xy-plane:

$$u = \frac{1+\nu}{8E} K_1 \sqrt{\frac{2r}{\pi}} \left[(2\kappa - 1)\cos\frac{\theta}{2} - \cos\frac{3\theta}{2} \right],$$ [5.8]

$$v = \frac{1+\nu}{8E} K_1 \sqrt{\frac{2r}{\pi}} \left[(2\kappa - 1)\sin\frac{\theta}{2} - \sin\frac{3\theta}{2} \right]$$

where for plane strain problems κ equals

$$\kappa = 3 - 4\nu$$ [5.9]

while for plane stress problems κ becomes

$$\kappa = \frac{3-\nu}{1+\nu}$$ [5.10]

Also applicable to plane strain problems, the displacements in z-direction are $w = 0$.

Mode II

Stresses in xy-plane:

$$\sigma_x = \frac{K_{II}}{\sqrt{2\pi r}} \cos\frac{\theta}{2} \left(2 - \cos\frac{\theta}{2}\cos\frac{3\theta}{2} \right),$$ [5.11]

$$\sigma_y = \frac{K_{II}}{\sqrt{2\pi r}} \sin\frac{\theta}{2}\cos\frac{\theta}{2}\cos\frac{3\theta}{2},$$

$$\tau_{xy} = \frac{K_{II}}{\sqrt{2\pi r}} \cos\frac{\theta}{2} \left(1 - \sin\frac{\theta}{2}\sin\frac{3\theta}{2} \right)$$

Displacements in xy-plane:

$$u = \frac{1+\nu}{8E} K_{II} \sqrt{\frac{2r}{\pi}} \left[(2\kappa + 3)\sin\frac{\theta}{2} - \sin\frac{3\theta}{2} \right],$$ [5.12]

$$v = \frac{1+\nu}{8E} K_{II} \sqrt{\frac{2r}{\pi}} \left[(2\kappa - 3)\cos\frac{\theta}{2} - \cos\frac{3\theta}{2} \right]$$

For plane strain problems, Equations (5.11) and (5.12) are supplemented by Equations (5.6) and (5.9). For plane stress problems they are supplemented by Equations (5.7) and (5.10) correspondingly.

Mode III
  Stresses:

$$\tau_{xy} = \frac{K_{III}}{\sqrt{2\pi r}} \sin\frac{\theta}{2}, \tag{5.13}$$

$$\tau_{xz} = \frac{K_{III}}{\sqrt{2\pi r}} \cos\frac{\theta}{2},$$

$$\sigma_x = \sigma_y = \sigma_z = \tau_{xy} = 0$$

Displacements:

$$w = \frac{1+\nu}{E} K_{III} \sqrt{\frac{2r}{\pi}} \sin\frac{\theta}{2}, \tag{5.14}$$
$$u = v = 0$$

Note: Mode III is an exception as far as the restriction of plane stress or plane strain is concerned. Due to the geometric form of the crack it does not apply here.

Stress intensity factor

Let us return to the stress intensity factor K introduced in Equation (5.4). As mentioned, it presents a relationship of a loading mode, geometry of the stressed part and the length of the crack. Mode I, the perpendicular crack opening, covers the most common form of cracks caused by fatigue. The basic equation of the stress intensity factor for Mode I can be stated

$$K_I = f_I \sigma \sqrt{\pi a} \tag{5.15}$$

Figure 5.4 shows four standard specimens each with a different geometry: (a) plate with a center crack loaded in tension; (b) plate with an edge crack loaded in tension; (c) plate with double edge cracks loaded in tension; and (d) beam with an edge crack loaded in bending. Let us present the compliance function f_I for each one of these cases (see Sih, Reference 2).

Center crack loaded in tension

$$f_I = \sqrt{\sec\frac{\pi a}{2b}} \tag{5.16}$$

Edge crack loaded in tension

$$f_I = 1.12 - 0.231\frac{a}{b} + 10.55\left(\frac{a}{b}\right)^2 - 21.72\left(\frac{a}{b}\right)^3 + 30.39\left(\frac{a}{b}\right)^4 \tag{5.17}$$

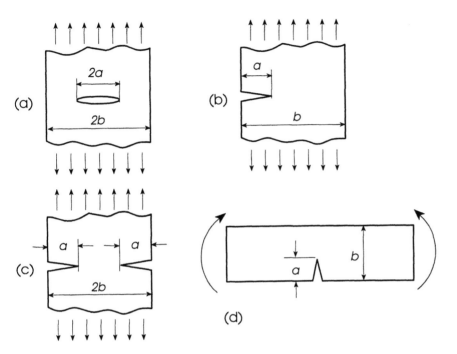

Figure 5.4 Concerning mode I: (a) center cracked plate in tension;
(b) edge cracked plate in tension; (c) double-edge cracked plate
in tension; (d) cracked beam in bending.

Double edge cracks loaded in tension

$$f_I = 1.12 + 0.203\frac{a}{b} - 1.197\left(\frac{a}{b}\right)^2 + 1.93\left(\frac{a}{b}\right)^3 \qquad [5.18]$$

Edge crack loaded in bending

$$f_I = 1.122 - 1.40\frac{a}{b} + 7.33\left(\frac{a}{b}\right)^2 - 13.08\left(\frac{a}{b}\right)^3 + 14.0\left(\frac{a}{b}\right)^4 \qquad [5.19]$$

For small cracks $(a/b << 1)$, we can simplify the above equations by approximations, converting the compliance functions as follows:

First, the term $\pi a/2b$ in Equation (5.16) becomes close to zero, therefore

$$f_I = 1.0 \qquad [5.20]$$

Second, we can ignore the higher order terms in Equations (5.17), (5.18) and (5.19), converting them into

$$f_I = 1.12 \qquad [5.21]$$

Sometimes it is convenient to present Equations (5.20) and (5.21) in a unified fashion $f_I = \alpha$. to express the stress intensity factor as

$$K_I = \alpha\sigma\sqrt{\pi a}$$ [5.22]

where α is a constant defined as follows:

$\quad \alpha = 1.0$ for a center crack loaded in tension,

$\quad \alpha = 1.12$ for an edge crack (or double cracks) loaded in tension, or an edge crack loaded in bending.

Similar expressions for small cracks ($a/b < < 1$) can be used for Modes II and III. For Mode II the stress intensity factor is expressed by

$$K_{II} = \tau\sqrt{\pi a} , \quad f_{II} = 1.0$$ [5.23]

and for Mode III it becomes

$$K_{III} = \tau\sqrt{\pi a} , \quad f_{III} = 1.0$$ [5.24]

Effect of notch on the stress intensity factor

Consider a notched plate under loading with an existing crack at the notch. See Figure 5.5. Since the geometry of the opening mode has changed, we must correct the expression of the stress intensity factor used heretofore. An empirical method is chosen to do this which takes into consideration the stress gradients in the vicinity of the notch. Let us take the simplified expression, Equation (5.22). As indicated in Figure 5.6, the area in the vicinity of the notch is separated between the area of high stress gradients

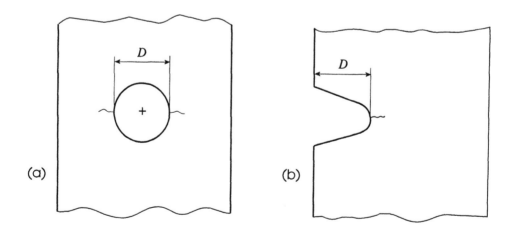

Figure 5.5 Cracks emanating from notches: (a) circular hole; (b) edge notch.

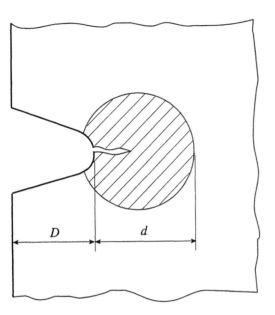

Figure 5.6 Stress field strongly influenced by the notch.

and the remaining field with low gradients. The extent of the stress field influenced strongly by the notch can be estimated according to Smith and Miller[3] as

$$d = 0.13\sqrt{Dr} \qquad\qquad [5.25]$$

Within this domain, the stress intensity factor is computed based on a stress concentration factor K_t. Thus K_I becomes

$$K_I = \alpha K_t S \sqrt{\pi a} \qquad\qquad [5.26]$$

In the remaining field, the stress intensity factor is assumed, per Dowling,[4] to equal

$$K_I = \alpha S \sqrt{\pi (D + a)} \qquad\qquad [5.27]$$

5.1.2 Griffith theory of fracture

Even though our main concern is the subject of crack propagation in elastic-plastic materials such as metals, it is necessary now to introduce a theory that applies to ideally brittle materials—the Griffith theory. Understanding the way this theory deals with the energy criteria within glass or ceramics will help us deal with more complex issues. Griffith's theory[5] explains the mechanical conditions of crack propagation in brittle materials where crystal imperfections grow into macroscopic size cracks under high stresses. The theory defines the bounds of elastic energy within which the crack propagation takes place.

Consider a cracked body as shown in Figure 5.7. The body is subject to external forces T_i. Let us analyze the equilibrium of the body while the surface of the crack increases by a virtual increment δA from the initial size A. Based on the principle of virtual displacement (Chapter 2), the equilibrium of three factors must be maintained— the elastic energy released during the crack propagation, the virtual work performed by external forces, and the change in total strain energy of the body.

Let us denote the elastic energy released during crack propagation by $\delta \Pi$. The other two factors are expressed as follows.

The virtual work performed by external forces

$$\delta W_F = \int_F T_i \delta u_i \delta s \qquad\qquad\qquad [5.28]$$

and the change in the total elastic strain energy of the body

$$\delta U = \int_{vol} \sigma_{ij} \delta\varepsilon_{ij} d(vol) = \int_{vol} \delta U_o \, d(vol) \qquad\qquad [5.29]$$

From the above, the equilibrium can be expressed as

$$\delta W_F - \delta \Pi - \delta U = 0 \qquad\qquad\qquad [5.30]$$

To express the Griffith criterion in mathematical form, we need to define the elastic energy released per unit of crack surface, G,

$$G = \frac{\delta \Pi}{\delta A} \qquad\qquad\qquad [5.31]$$

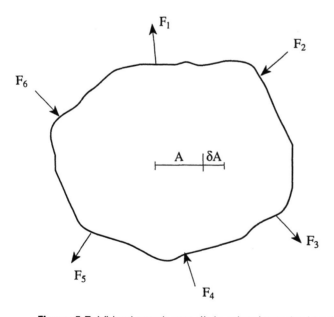

Figure 5.7 Virtual crack growth in a body under load.

Griffith formulated the condition for the crack propagation which is that the energy released must be sufficient for the destruction of material when G is equal to or exceeds a certain critical value G_c

$$G \geq G_c \qquad\qquad [5.32]$$

The critical value G_c is determined experimentally, examining fracturing of plates with existing cracks, as will be described later.

In addition to the above, the Griffith theory states that the mechanism of crack propagation is a function of the elastic strain energy alone and is independent of the type of load application. To explain this, let us compare two cases with different load applications as shown in Figure 5.8. In Case (a), the load is applied in the form of an outer force P, while in Case (b) the load is applied as a given displacement z.

Case (a)   Consider a plate of unit thickness with a crack of length a. See Figure 5.8(a). The plate is subjected to a tension force P and elongation z. (z is the displacement at the location of force P.) The elastic strain energy of the plate equals

$$U = \frac{1}{2} Pz \qquad\qquad [5.33]$$

Since displacement z is proportional to P, it can be expressed as

$$z = C\!\left(a\right)P \qquad\qquad [5.34]$$

$C(a)$ is the compliance of the plate and is a function of crack size a. Using the latter equation, the strain energy can be expressed as

$$U = \frac{1}{2}C\!\left(a\right)P^2 \qquad\qquad [5.35]$$

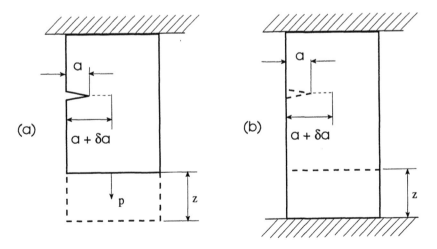

Figure 5.8 Concerning Griffith criterion: (a) crack propagation in an expanding body; (b) crack propagation in a fixed body.

or

$$U = \frac{1}{2} \frac{z^2}{C(a)} \qquad\qquad [5.36]$$

As the crack grows by a virtual increment δa, the elastic strain energy of the plate changes by

$$\delta U = \frac{\partial U}{\partial a} \delta a \qquad\qquad [5.37]$$

The virtual work performed by force P equals

$$\delta W_F = P \frac{\partial z}{\partial a} \delta a \qquad\qquad [5.38]$$

The condition of equilibrium, Equation (5.30) (where index I denotes Mode I of the crack), now becomes

$$P \frac{\partial z}{\partial a} \delta a - G_I \delta a - \frac{\partial U}{\partial a} \delta a = 0 \qquad\qquad [5.39]$$

This Equation (5.39), by means of Equations (5.34) and (5.35), converts the expression of the energy release rate, G_I, for this case

$$G_I = P^2 \frac{\partial C}{\partial a} + CP \frac{\partial P}{\partial a} \delta a - \frac{1}{2} P^2 \frac{\partial C}{\partial a} - CP \frac{\partial P}{\partial a} \qquad\qquad [5.40]$$

$$= \frac{1}{2} P^2 \frac{\partial C}{\partial a} = \frac{1}{2} \frac{z^2}{C^2} \frac{\partial C}{\partial a}$$

Case (b)   Consider now an undamaged plate of the same dimensions as in Case (a), stretched by distance z and then fixed at its ends. See Figure 5.8(b). Crack a is introduced by an outside operation. Subsequently, due to instability, the crack propagates by itself, growing by increment δa until equilibrium is reached. Because the work of external forces, δW_F, is zero, the condition of equilibrium, Equation (5.30), has the form

$$G_I \delta a + \frac{\partial U}{\partial a} \delta a = 0 \qquad\qquad [5.41]$$

From here, by means of Equation (5.36), the expression of the energy release rate, G_I, for this case becomes

$$G_I = -\frac{\partial U}{\partial a} = \frac{1}{2} \frac{z^2}{C^2} \frac{\partial C}{\partial a} \qquad\qquad [5.42]$$

The above proves that the energy release rate G_I in Equation (5.42) is the same as in Equation (5.40), that G_I is independent of the type of load application, and that it may be determined by the strain energy alone.

Energy release rate

We shall develop the expression of energy release rate G_I based on strain energy as per Rice and Drucker.[6] It is assumed that the strain energy that was released during the crack growth equals the work that is required to force it to close. Consider again an infinite plate containing a crack. See Figure 5.9. We shall evaluate the work along the segment $a < x < a + \delta a$ on $0-x$ axis between the two instances (a) and (b) shown in the figure. In the open condition, Figure 5.9(a), the stresses and displacements along the segment equal

$$\sigma_y(r') = 0 \qquad [5.43]$$

$$v(r') = \frac{1+v}{2E} K_I \sqrt{\frac{2r'}{\pi}} (\kappa + 1) \qquad [5.44]$$

where r' is a point on the segment defined by

$$r' = a + \delta a - x \qquad [5.45]$$

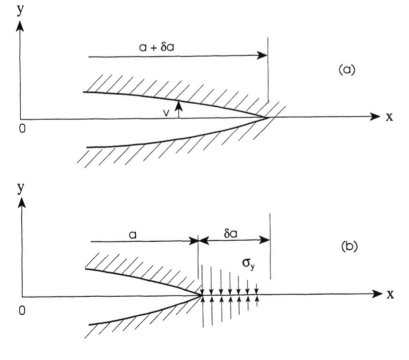

Figure 5.9 Closure of a crack.

In the closed position, Figure 5.9(b), the stresses and displacements become

$$\sigma_y(r) = \frac{K_I}{\sqrt{2\pi r}} \tag{5.46}$$

$$v(r) = 0 \tag{5.47}$$

where r is defined by

$$r = x - a \tag{5.48}$$

The strain energy of the crack in closed position equals

$$\delta U = -2 \cdot \frac{1}{2} \int_a^{a+\delta a} \sigma(r) v(r') dx \tag{5.49}$$

$$= -\frac{1+v}{E} \frac{K_I^2}{2\pi}(\kappa + 1) \int_a^{a+\delta a} \sqrt{\frac{a + \delta a - x}{x - a}} dx$$

Now the right-hand integral in Equation (5.49) is

$$\int_a^{a+\delta a} \sqrt{\frac{a + \delta a - x}{x - a}} dx = \int_a^{a+\delta a} \sqrt{\frac{\delta a}{x - a} - 1} \, dx = \frac{\pi \delta a}{2} \tag{5.50}$$

whereby the strain energy becomes

$$\delta U = -\frac{1}{4} \frac{1+v}{E} K_I^2 (\kappa + 1) \delta a \tag{5.51}$$

It follows that the energy release rate is a function of the stress intensity factor and it equals

$$G_I = -\frac{\delta U}{\delta a} = \frac{1}{4} \frac{1+v}{E}(\kappa + 1) K_I^2 \tag{5.52}$$

For plane stress condition the energy release rate, by virtue of Equation (5.10), can be expressed as

$$G_I = \frac{K_I^2}{E} \tag{5.53}$$

while for plane strain condition, based on Equation (5.9), the energy release rate equals

$$G_I = (1 - v^2) \frac{K_I^2}{E} \tag{5.54}$$

The above was proven for Mode I, the opening mode. It can be shown that for Mode II, the sliding mode, similarly, we get for plane stress condition

$$G_{II} = \frac{K_{II}^2}{E} \quad\quad\quad\quad [5.55]$$

and for plane strain condition

$$G_{II} = \left(1 - v^2\right)\frac{K_{II}^2}{E} \quad\quad\quad\quad [5.56]$$

For Mode III, the tearing mode, we get

$$G_{III} = \left(1 - v^2\right)\frac{K_{III}^2}{E} \qu\quad\quad\quad [5.57]$$

Fracture toughness

As noted above, at the instance of instability when the final fracture takes place, the energy release rate must reach a critical value. In view of the above expressions, however, it becomes apparent that the instance of instability can be expressed by either one of the relevant material properties—the critical energy release rate G_c or the critical stress intensity factor K_c, called the fracture toughness. The fracture toughness can be determined experimentally from tests with predetermined crack size a.

Figure 5.10 presents two standard test specimens that are used to measure the fracture toughness of metals—specimens (a) and (b). One can compute the fracture toughness from test data by means of the equation

$$K_{Ic} = f_I \sigma_c \sqrt{\pi a} \quad\quad\quad\quad [5.58]$$

The fracture toughness for specimen (a) as per the above equation becomes

$$K_{Ic} = \frac{PS}{BW^{1.5}}\left[2.9\left(\frac{a}{W}\right)^{0.5} - 4.6\left(\frac{a}{W}\right)^{1.5} + 21.8\left(\frac{a}{W}\right)^{2.5} - 37.6\left(\frac{a}{W}\right)^{3.5} + 38.7\left(\frac{a}{W}\right)^{4.5}\right] \quad [5.59]$$

and for specimen (b) it equals

$$K_{Ic} = \frac{PS}{BW^{1.5}}\left[29.6\left(\frac{a}{W}\right)^{0.5} - 185.5\left(\frac{a}{W}\right)^{1.5} + 655.7\left(\frac{a}{W}\right)^{2.5} - 1017\left(\frac{a}{W}\right)^{3.5} + 638.9\left(\frac{a}{W}\right)^{4.5}\right] \quad [5.60]$$

5.1.3 Extension of LEFM into the plastic domain

Before we make the transition to include the process of crack propagation in plastic conditions, let us reiterate the dictum of previous discussions: the critical energy release

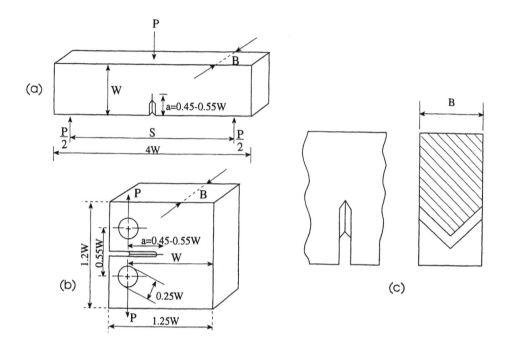

Figure 5.10 Standard test specimens: (a) bend specimen; (b) compact tension specimen; (c) chevron notch.

rate G_c must be sufficient to overcome the resistance to fracture. The resistance to fracture involves two factors—γ_s, the surface energy to initiate new surfaces and γ_p, the energy to form a plastic zone in the vicinity of the crack. See Griffith (Reference 5) and Orowan.[7] Empirically, a correlation of the two factors can be expressed by

$$G_c = 2\left(\gamma_s + \gamma_p\right) \tag{5.61}$$

For truly brittle materials, such as glass or ceramics, term γ_p equals zero. For metals and polymers, however, the plastic energy predominates and γ_p is greater than γ_s by several orders of magnitude. Consequently, ignoring plasticity and continuing with an assumption of linear elasticity throughout the material will cause an inherent error in the solutions presented for metal parts.

Energy release rate and J integral

Now let us concentrate on the elastic-plastic behavior of metals in the vicinity of crack and review the energy release rate true for plastic conditions. We begin with the condition of equilibrium. Consider the body with a crack under load in Figure 5.7. The energy released during crack propagation, $\delta\Pi$, equals

$$\delta\Pi = \delta\int_{\text{vol}} U_o \, d(\text{vol}) - \delta\int_\Gamma T_i u_i \, ds \tag{5.62}$$

In this equation, the first integral refers to the volume of the body while the second integral refers to its circumferential surface Γ. Here we substitute for the elastic energy release rate G a term that refers to the elastic-plastic condition, J. The elastic-plastic energy release is expressed by

$$J = -\frac{d\Pi}{dA} = -\frac{d}{dA}\left[\int_{vol} U_o \, d(vol) - \int_\Gamma T_i u_i \, ds\right] \tag{5.63}$$

The above Equation (5.63) generally applies to a three-dimensional space. For the purpose of discussing the energy release rate in two-dimensional body of unit thickness, the latter equation is revised as follows

$$J = -\frac{d}{dA}\left(\int_{vol} U_o \, dxdy - \int_\Gamma T_i u_i \, ds\right) \tag{5.64}$$

By virtue of theorem of energy conservation,[8] the right-hand expression becomes a path integral

$$J = \int_\Gamma \left(U_o dy - T_i \frac{\partial u_i}{\partial x_i} ds\right) \tag{5.65}$$

Let us consider a part of the plate bounded by a contour Γ_1. See Figure 5.11. The applied load is in a form of traction forces

$$T_i = \sigma_{ij} n_j \tag{5.66}$$

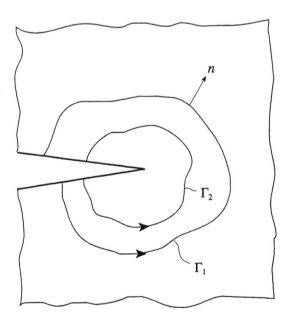

Figure 5.11 J-contour around crack tip.

where σ_{ij} denotes the stresses along contour Γ and n_j is a normal to Γ. Thus, for example, if we chose a circular contour (see Figure 5.12) Equation (5.65) becomes

$$J = \int_{-\pi}^{\pi} \left(U_o \cos\theta - T_i \, \frac{\partial u_i}{\partial x_i} \right) r d\theta \qquad [5.67]$$

It was shown by Rice (Reference 8) that in both linear-elastic and nonlinear-elastic problems the integral defined by Equation (5.65) does not depend on Γ, that is, it is path-independent. Although there is no theoretical proof, the integral is also assumed to be path-independent for elastic-plastic conditions. The integral J thus represents the elastic-plastic energy release rate. It was confirmed experimentally (for thick specimens)[9] that a critical elastic-plastic energy release rate J_c, similar to G_c, exists. In cases where no plastic deformation takes place (see Begley and Landes, Reference 9) the critical energy release rate for linear elastic plane-strain cracks in the opening mode becomes

$$J_{Ic} = G_{Ic} = \left(1 - v^2\right) \frac{K_{Ic}^2}{E} \qquad [5.68]$$

Plastic deformation and fracture toughness

As in the case of linear elasticity, the instance of instability for the elastic-plastic conditions may be expressed by either the critical energy release rate J_c, or fracture toughness K_c. Let us review a method which with the help of approximation can be used to determine the fracture toughness for elastic-plastic materials. The method is shown schematically in Figure 5.13, as per Irwin.[10] In the figure a cracked plate in Mode I is

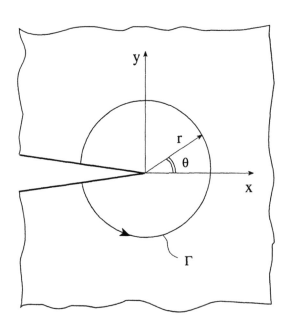

Figure 5.12 Circular J-contour around crack tip.

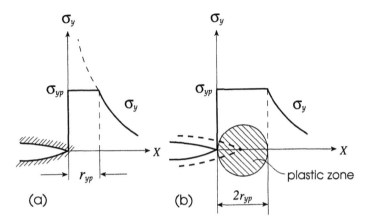

Figure 5.13 Plastic zone at crack tip: (a) definition of size of plastic zone; (b) corrected crack tip model.

presented subject to plane stress conditions. Consider the stress distribution in the vicinity of the crack, as shown in Figure 5.13(a). According to the elastic solution, the stress σ_y along $0x$-axis equals

$$\sigma_y = \frac{K_I}{\sqrt{2\pi r}}$$

[5.69]

In the domain $r < r_{yp}$ where σ_y reaches the elastic limit σ_{yp}, we assume a constant stress distribution. The location r_{yp} can be computed from

$$r_{yp} = \frac{1}{2\pi}\left(\frac{K_I}{\sigma_{yp}}\right)^2$$

[5.70]

Consequently, considering the effect of plasticity, the stress distribution is corrected as follows

$$\sigma_y = \sigma_{yp} \quad \text{at} \quad 0 < r < r_{yp}$$

[5.71]

$$\sigma_y = \frac{K_I}{\sqrt{2\pi r}} \quad \text{at} \quad r > r_{yp}$$

[5.72]

as shown in the figure. Irwin's method assumes a theoretical crack whose length equals

$$a_{eff} = a + r_{yp}$$

[5.73]

See Figure 5.13(b). It follows that the stress intensity factor has to be revised correspondingly to become

$$K_{I,\text{eff}} = \alpha\sigma\sqrt{\pi\left(a + r_{yp}\right)} \tag{5.74}$$

The computation of the revised stress intensity factor $K_{I,\text{eff}}$ involves an iterative procedure using Equations (5.70) and (5.74). The resulting stress distribution is shown in Figure 5.13(b).

Equation (5.70) was found valid for plane stress conditions only. For plane strain Irwin introduced (instead of σ_{yp}) an effective yield point

$$\sigma_{yp,\text{eff}} = 1.68\sigma_{yp} \tag{5.75}$$

Therefore for plane strain the dimension r_{yp}, instead of Equation (5.70), is defined by

$$r_{yp} = \frac{1}{6\pi}\left(\frac{K_I}{\sigma_{yp}}\right)^2 \tag{5.76}$$

Experimenting with metals for fracture toughness, one finds that the results are influenced by the thickness of the specimen, crack size, and other factors. For instance in testing thin specimens where the plane stress condition exists, the plastic zone is large as confirmed by Equation (5.70). On the other hand, in testing thick specimens where the plane strain prevails, the plastic zone is smaller, see Equation (5.76). Consequently the effective fracture toughness K_{Ic} in the plane stress condition is larger than the one in the plane strain condition (known to be from 2 to 10 times larger), as follows from Equation (5.74).

In tests of specimens with different thicknesses one notes that K_{Ic} decreases asymptotically as the specimen's thickness increases because of a smaller plastic zone. In test procedures for fracture toughness a certain thickness must be maintained in order to obtain the condition of the plane strain. Only then can a limit value of fracture toughness be reached which is used as the basic value accepted in standard test procedures. See Figure 5.14. In fact, it has been estimated empirically that for plane strain condition the minimum material thickness B must be

$$B \geq 2.5\left(\frac{K_{Ic}}{\sigma_{yp}}\right)^2 \tag{5.77}$$

See Brown and Srawley.[11]

R curve The thickness of a specimen is a predominant factor in assessing the limits of stability set by fracture toughness. For thick plates we relied on the relationship between the stress intensity factor and fracture toughness to define the onset of instability—a final fracture takes place when the stress intensity factor reaches the value of fracture toughness of the material. Considering thin plates under plane stress, we are faced with a problem—the existence of plane stress conditions makes the limits of instability variable because of plasticity. To determine the limits of instability we use the R curve.

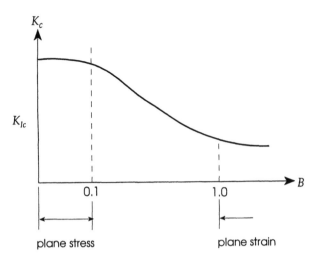

Figure 5.14 Effect of plate thickness on fracture toughness.

Let us denote the surface resistance to crack propagation by R. As per Griffith criterion, at the onset of instability the energy release rate G must exceed the surface resistance R. The crack resistance R in plane strain conditions is a function of the fracture toughness which is a constant. In the case of plane stress conditions, R varies as the crack propagates, requiring additional information such as the specimen thickness and crack length.

We concentrate on the dependence of two factors—surface resistance and length, R versus a. This is shown graphically in Figure 5.15. In the figure, the straight lines starting at coordinate origin O correspond to the function (Mode I is assumed)

$$G_{\mathrm{I}} = \frac{\pi \sigma^2 a}{E} \qquad [5.78]$$

Figure 5.15 R-curve in a G-a diagram.

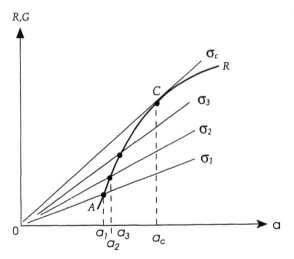

Suppose a cracked plate having a crack of size a_1 loaded to stress σ_1. Intersection point A denotes that the energy release rate equals surface resistance. If stress σ stays constant, the propagation cannot exist beyond point A, as confirmed by the fact that the G line is below the R curve. If σ is getting larger, however, the crack grows through sizes a_2, a_3, etc., consecutively, until it reaches size a_c which corresponds to the tangent point C. Point C is defined by the conditions

$$\frac{\partial G}{\partial a} = \frac{\partial R}{\partial a} \qquad\qquad [5.79]$$

and

$$G = R \qquad\qquad [5.80]$$

Beyond point C the G line is above the R curve, revealing a condition of instability and that is when a final fracture takes place.[12] The energy release rate at point C equals $G = G_c$. The value of K_c, corresponding to this point, is referred to as the plane-stress fracture toughness. It should be noted that the application of the above method to determine the plane-stress fracture toughness requires sufficient experimental data.

Crack opening displacement

A known method to determine the fracture toughness in elastic-plastic conditions is the crack-opening displacement (COD) method.[13] The application of the method concentrates on the tip of the crack. Figure 5.16 shows a model of a crack in tension with the crack-opening displacement δ. Displacement δ is represented by the approximate expression

$$\delta = \frac{\pi \sigma^2 a}{E \sigma_{yp}} \qquad\qquad [5.81]$$

At the onset of instability $\delta = \delta_c$ which allows us to derive the fracture-toughness correlation

$$\frac{\delta_c}{\varepsilon_{yp}} = \left(\frac{K_{Ic}}{\sigma_{yp}} \right)^2 \qquad\qquad [5.82]$$

The application of the COD method requires an experimental determination of parameter δ_c, a material property.

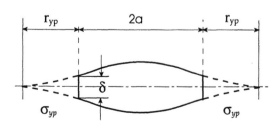

Figure 5.16 Crack opening displacement (COD) model.

5.2
CRACK PROPAGATION UNDER FATIGUE LOAD

Up to now the discussion of crack propagation has been concerned with steady or monotonic loading, not taking into consideration the fatigue caused by fluctuating loading. Thus we considered the instance of failure as culmination of a steady crack growth. Experience shows that in metal parts subject to fluctuating loading the process is uneven and includes a preparatory phase—during which the crack growth is sub-critical—and the critical phase leading to fracture. This was observed in Figure 5.1. There a broken part due to fatigue has a different surface up to the rupture zone, signalling that the stress intensity factor reached the critical value. Figure 5.17 presents the history of crack growth under a fluctuating loading. Attention is brought to the fact that the crack grows exponentially, slowly at first, with the rate of propagation increasing continuously. The final failure, point C, occurs when the crack size reaches a_c and the stress intensity factor reaches the critical value K_c. In Figure 5.17 the fatigue histories of two specimens subject to different stress ranges are compared. One notes that the specimen subject to the higher stress range reaches the critical point earlier, failing with a crack of a smaller size.

We will follow the discussion of crack growth considering it as a function of stress intensity factor K. We begin the analysis with fatigue loads fluctuating at a constant amplitude and extend it to more complex problems of fatigue loads with a variable amplitude.

5.2.1 Constant amplitude loading

Consider a fatigue load that fluctuates at a constant amplitude where the stresses vary between constant limits S_{max} and S_{min}. The range of the stress intensity factor can be expressed as

$$\Delta K = K_{max} - K_{min} = \alpha\left(S_{max} - S_{min}\right)\sqrt{\pi a} \qquad [5.83]$$

To continue developing a mathematical formulation of crack propagation, we refer to typical test data[14] shown in Figure 5.18. The figure presents the crack growth rate per

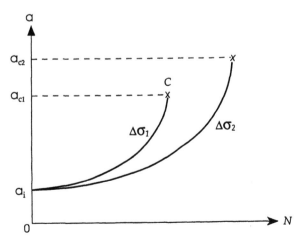

Figure 5.17 Fatigue crack propagation at constant stress range $\Delta\sigma$. Note: $\Delta\sigma_1 > \Delta\sigma_2$.

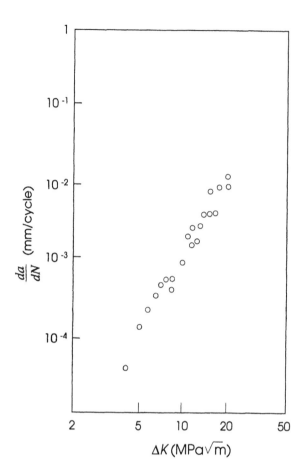

Figure 5.18 Fatigue crack growth versus stress intensity factor for aluminum 7075-T6, $S_{min} = 0$. (Data from Hudson, C. M., see Reference 14.)

cycle plotted versus ΔK. A careful study of the results shows that the data fit into a narrow band which can be approximated by a sigmoidal curve as shown in Figure 5.19. The curve can be divided into three distinct sections that represent different regions of crack growth: region I, crack formation; region II, moderate crack propagation; and region III, accelerated crack growth and fracture. Point ΔK_{th} denotes a threshold value, the onset of the crack growth. ΔK_c reflects the point of final failure. Under consideration is the region II.

As per Paris, Gomez and Anderson,[15] the crack growth can be expressed in the form of

$$\frac{da}{dN} = f(\Delta K) \tag{5.84}$$

Indeed, the test data in region II show a linear correlation

$$\frac{da}{dN} = C(\Delta K)^n \tag{5.85}$$

where C is a constant factor and n an exponent taken from tests. The above equation, first presented by Paris and Endoyan,[16] is known as the fatigue crack propagation law.

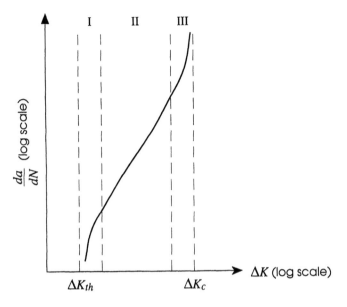

Figure 5.19 Three regions of crack growth rate.

To compute the life expectancy of a cracked part we use a two-step approach. Equation (5.85) is inverted

$$dN = \frac{da}{C(\Delta K)^n}$$

[5.86]

and the latter is integrated to get the expression (valid for $n \neq 2$)

$$N_f = \frac{1}{C} \int_{a_i}^{a_f} \frac{da}{(\Delta K)^n} = \frac{1}{C\left[\alpha(S_{max} - S_{min})\sqrt{\pi} \right]^n} \int_{a_i}^{a_f} \frac{da}{a^{n/2}}$$

[5.87]

$$= \frac{1}{(n-2)C\left[\alpha(S_{max} - S_{min})\sqrt{\pi} \right]^n} \left[\frac{1}{a_i^{(n-2)/2}} - \frac{1}{a_f^{(n-2)/2}} \right]$$

a_i is the initial crack length computed from the equation

$$a = \frac{1}{\pi} \left[\frac{\Delta K}{\alpha(S_{max} - S_{min})} \right]^2$$

[5.88]

using

$$\Delta K_{th} = K_{th} - K_{min}$$

[5.89]

a_f is the crack length at failure computed from Equation (5.88), on the basis of

$$\Delta K_c = K_c - K_{min} \qquad\qquad [5.90]$$

Equation (5.89) is superfluous in instances where the life expectancy is questioned for a part with existing signs of fatigue—the presence of a crack with a given size.

As can be seen from the above, the two source data (both obtained experimentally) enabled us to arrive at the number of fluctuations to failure, that is, the stress intensity factor at threshold point K_{th} and the fracture toughness K_c. The accumulated test data of these two factors provide the values for practical application in design for fatigue. Values of K_{th} and K_c for different metals are listed in Appendix.

Effect of S_{min}

Let us examine now how the life expectancy analysis is influenced by different fatigue loadings at constant amplitude but with a changing lower stress limit. The three cases that follow review the life expectancy with S_{min} values positive, negative, or equaling zero. See Figure 5.20.

Case (a) $S_{min} = 0$ For this case see Figure 5.20(a). The range of stress intensity factor is

$$\Delta K = K_{max} = \alpha S_{max} \sqrt{\pi a} \qquad\qquad [5.91]$$

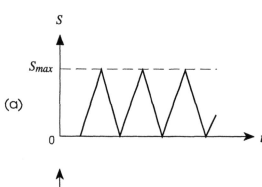

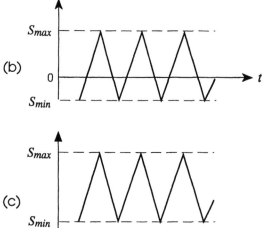

Figure 5.20 Three cases of fatigue loading at constant amplitude.

causing a change in the life expectancy expression, Equation (5.87), as

$$N_f = \frac{1}{C\left(\alpha S_{max}\sqrt{\pi}\right)^n} \int_{a_i}^{a_f} \frac{da}{a^{n/2}}$$

[5.92]

$$= \frac{1}{(n-2)C\left(\alpha S_{max}\sqrt{\pi}\right)^n} \left[\frac{1}{a_i^{(n-2)/2}} - \frac{1}{a_f^{(n-2)/2}} \right]$$

Thus we arrived at the number of fluctuations to failure when $S_{min} = 0$. As per Barsom,[17] experimentally obtained factor C and exponent n for different types of steel are as follows.

Ferritic-pearlitic steel

$$\frac{da}{dN}\left(\text{m/cycle}\right) = 6.9 \times 10^{-12} \times \left(\Delta K \ \text{MPa}\sqrt{\text{m}}\right)^{3.0},$$

[a]

$$\frac{da}{dN}\left(\text{in/cycle}\right) = 3.6 \times 10^{-10} \times \left(\Delta K \ \text{ksi}\sqrt{\text{in}}\right)^{3.0}$$

Martensitic steel

$$\frac{da}{dN}\left(\text{m/cycle}\right) = 1.35 \times 10^{-10} \times \left(\Delta K \ \text{MPa}\sqrt{\text{m}}\right)^{2.25},$$

[b]

$$\frac{da}{dN}\left(\text{in/cycle}\right) = 6.6 \times 10^{-9} \times \left(\Delta K \ \text{ksi}\sqrt{\text{in}}\right)^{2.25}$$

Austenitic steel

$$\frac{da}{dN}\left(\text{m/cycle}\right) = 5.6 \times 10^{-12} \times \left(\Delta K \ \text{MPa}\sqrt{\text{m}}\right)^{3.25},$$

[c]

$$\frac{da}{dN}\left(\text{in/cycle}\right) = 3.0 \times 10^{-10} \times \left(\Delta K \ \text{ksi}\sqrt{\text{in}}\right)^{3.25}$$

Case (b) $S_{min} < 0$ See Figure 5.20(b). Condition of compression at the lower limit reflects the fact that the crack stops growing in this condition (the propagation occurs at tensional stresses only). Therefore we may ignore S_{min} in Equation (5.87) and apply here the same equation as in Case (a) above, Equation (5.92).

Case (c) $S_{min} > 0$ See Figure 5.20(c). Condition of tension at the lower limit, necessitates a recall of the basic crack growth correlation, Equation (5.84). To apply it here, a correction for S_{min} is needed after which the basic correlation becomes

$$\frac{da}{dN} = f(\Delta K, R)$$

[5.93]

where R is the stress ratio defined by

$$R = \frac{S_{min}}{S_{max}}$$

[5.94]

Further, the fatigue crack propagation law, Equation (5.85) must be amended considering the test data (as shown in Figure 5.21). There are a number of empirical correlations to fit the amended fatigue crack propagation law. One, developed by Forman, Kearney and Engle,[18] has the form

$$\frac{da}{dN} = \frac{A(\Delta K)^n}{(1-R)K_c - \Delta K}$$

[5.95]

where A and n are material properties.

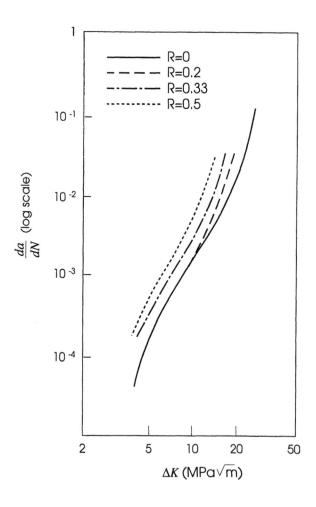

Figure 5.21 The influence of stress ratio on fatigue growth rate for aluminum alloy 7075-T6. (Data from Hudson, C. M., see Reference 14.)

Since the above presents a more cumbersome integration effort, let us consider a simpler approach based on the same test data (see Hudson, Reference 14) correlated differently as follows

$$\frac{da}{dN} = f\left(K_{max}, R\right) \tag{5.96}$$

Refer to Figure 5.22 with test data (Hudson, Reference 14) forming a family of sigmoidal curves where the left curve corresponds to $S_{min} = 0$, as in Case (a). One must be aware that the application of this curve in Case (c) with stress ratios $R > 0$, introduces an error. Since the error points toward more conservative estimations, which for design purposes provides safer limits, the resulting life expectancy is relevant in practice.

In practice the problems stated in Cases (a), (b) and (c) are solved numerically with the help of digital computers. The following presents a basic computational procedure:[19]

1. Assume an initial crack size a_i and an increment of crack growth Δa_i.

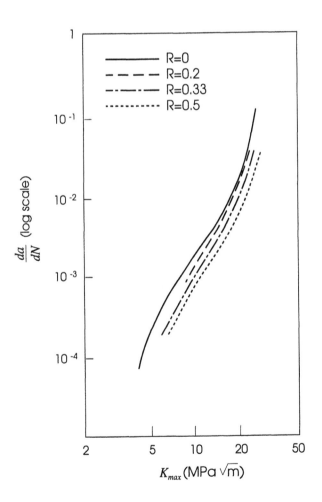

Figure 5.22
Crack propagation as function of K_{max} for aluminum alloy 7075-T6. (Data from Hudson, C. M., see Reference 14.)

2. Determine average ΔK.
3. Compute ΔN_i using available data

$$\frac{\Delta a_i}{\Delta N_i} = f\left(\Delta K, R\right)$$

4. Compute next step $a_{i+1} = a_i + \Delta a_i$ and repeat steps 1 to 3 until a_f is reached.
5. Compute $N = \Sigma(\Delta N_i)$.

5.2.2 Variable amplitude loading

We can now extend the above approach to problems with spectrum loading where the stress amplitude is not constant. There are several applicable methods, two of which are presented here: the block method and the statistical method. We will also mention physical phenomena caused the varable amplitude, such as crack retardation due to occasional overloading.

Block method   The method is based on arranging the random history of load fluctuations in terms of blocks. (See Leis, Nelson and Socie, Reference 19.) Since this method is presented in Chapter 3 (Section 3.4.2) and is identical in its application here, we will skip the preliminary explanations and present the crack growth in terms of blocks, as follows:

$$\Delta N_B = \sum_{k-1}^{k} \frac{\Delta a_i}{f\left(\Delta K_i, R_i\right)} \qquad [5.97]$$

Upon summation Equation (5.97) provides the final life in terms of number of fluctuations to failure

$$N = \Sigma\, \Delta N_B \qquad [5.98]$$

(The numerical computation of the life expectancy by this method is usually done with the help of a digital computer.)

Statistical method   The basis for this method is an observation by Barsom[20] that under random loading the rates of crack growth are approximately the same as under constant amplitude loading. If so, one can introduce a statistical root-mean-square intensity factor based on the equation

$$K_{\mathrm{rms}} = \sqrt{\frac{\sum\limits_{i=1}^{k}\left(\Delta K_i\right)^2}{k}} \qquad [5.99]$$

where k is the number of (ΔS, S_m) conditions per block. The equation allows us to compute the average fatigue crack growth rate using the correlation

$$\frac{\mathrm{d}a}{\mathrm{d}N} = C\left(\Delta K_{\mathrm{rms}}\right)^n \qquad [5.100]$$

From here the computation of life expectancy continues in the manner shown previously.

Retardation effect We extend our discussion to problems that include the phenomenon of occasional tensile overloading, that is, intermittent superimposed load jumps exceeding the regular upper limits. Experiments show that when to the regular loads, fluctuating at a constant amplitude, are added occasional overloads the crack propagation is retarded. This can be seen from a comparison of histories of crack growth in a specimen under fluctuating load, at a constant amplitude, with one experiencing intermittent overloads. See Figure 5.23. We find that the specimen subjected to intermittent overloads fails later. This can be explained by the fact that the overload causes the tensile stresses at the crack tip to exceed the yield point, creating an overload plastic region. See Figure 5.24. Following the cycle of the overload, residual compressive stresses in the plastic region tend to close the crack, retarding the crack propagation. The crack growth resumes after the residual stresses are overcome by the subsequent cycles.

There are several mathematical models to analyze the process of retardation. Presented here is one of the models per Wheeler.[21] The model is based on an integration of the crack length, cycle-by-cycle. The basic correlation of the crack growth remains the same as under the constant amplitude loading, Equations (5.84) or (5.93). The crack length after a number of applications equals the sum as expressed by

$$a_k = a_0 + \sum_{i=1}^{k} \Delta a_i = a_0 + \sum_{i=1}^{k} A_i f\left(\Delta K_i, R\right) \cdot \Delta N_i \qquad [5.101]$$

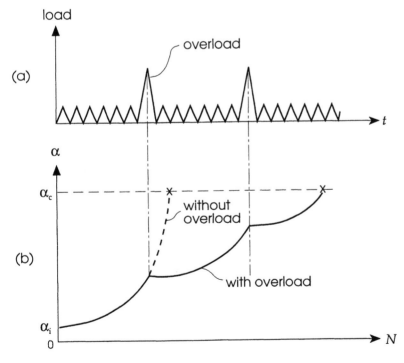

Figure 5.23 Overload effect on fatigue crack: (a) loading history; (b) crack growth with intermittent overloads.

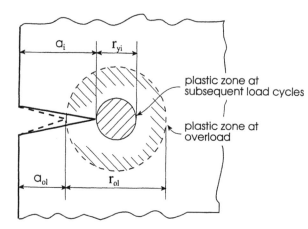

Figure 5.24 Retardation of crack growth.

where a_0 is the original crack length and A_i is a retardation factor due to the overload. The retardation factor is explained with the help of Figure 5.24 where length r_{0L} designates the size of the plastic zone after occurrence of the overload, while length a_{0L} shows the corresponding crack size. At a subsequent loading cycle (within the regular limits) the size of the new plastic zone would be r_{ypi}. and the crack size a_i. If this plastic zone falls within the overload plastic region, expressed by the condition

$$a_i + r_{yp,i} < a_{0L} + r_{0L} \qquad\qquad\qquad [5.102]$$

then the retardation factor equals

$$A_i = \left[\frac{r_{yp,i}}{\left(a_{0L} + r_{0L}\right) - a_i} \right]^m \qquad\qquad\qquad [5.103]$$

where exponent m is an empirical parameter. For steel generally $m = 1.3$. If a subsequent plastic zone is out of the overload plastic region, the new condition is

$$a_i + r_{yp,i} \geq a_{0L} + r_{0L} \qquad\qquad\qquad [5.104]$$

and there is no retardation, that is, $A_i = 1$.

The computation of the crack growth as per Equation (5.101) is usually done with the help of a digital computer. The computational procedure comprises the following steps:[22]

1. Assume an initial crack size a_i and an increment of crack growth Δa_i.
2. Compute average ΔK_i and r_{ypi}.
3. Check if $a_i + r_{yi} < a_{0L} + r_{0L}$ (whether yes or no) and determine A_i.
4. Compute ΔN_i retarded using available data

$$\frac{\Delta a_i}{\Delta N_i} = f\left(\Delta K, R\right)$$

5. Compute next step $a_{i+1} = a_i + \Delta a_i$ and repeat steps 1 to 4 until a_f is reached.
6. Compute $N = \Sigma(\Delta N_i)$.

5.3
MIXED LOADING AND MULTIAXIAL EFFECTS

As mentioned at the beginning of this chapter, the basic theory of crack propagation refers to two-dimensional elasticity theory. The problems reviewed heretofore have been concerned with one-dimensional loadings within plane space, ignoring in the solutions mixed loading, crack propagation changing directions and multiaxial effects. In recent years substantial effort has been directed toward consideration of such complex problems and toward development of working correlations applicable to multiaxial fatigue design including mixed loading problems. In this section we shall discuss some theoretical and empirical correlations, restricting the discussion to two types of phenomena: plane cracks under mixed loading and three-dimensional cracks under one-dimensional loading.

5.3.1 Plane cracks under mixed loading

By mixed loading we mean a combination of different modes of one-dimensional loadings applied simultaneously in plane space. The approach to analysis requires separate consideration of loading modes, deriving correlations by superposition of individual solutions.

To illustrate the approach, we consider a plate with a plane crack subjected to mixed loadings of Modes I, II and III. According to linear elasticity, the combined stresses will be

$$\sigma_{ij} = \frac{K_I}{\sqrt{2\pi r}} f_{Iij}(\theta) + \frac{K_{II}}{\sqrt{2\pi r}} f_{IIij}(\theta) + \frac{K_{III}}{\sqrt{2\pi r}} f_{IIIij}(\theta) \qquad [5.105]$$

The total energy release rate for all three modes together equals the sum

$$G_t = G_I + G_{II} + G_{III} \qquad [5.106]$$

For a crack to propagate, we know, the total energy release rate must reach a critical value, G_{tc}. In the analysis of mixed loading an additional factor plays a major role—the direction of the crack propagation. As an example, we analyze a crack subject to two modes—Mode I, opening and Mode II, sliding—resulting in so-called inclined crack.

Example: Inclined crack   The inclined crack is shown in Figure 5.25. Based on the given combined loading, we shall arrive at the direction of crack propagation and at the critical value of energy release rate for both plane strain and plane stress conditions.

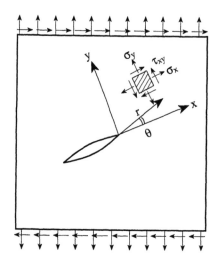

Figure 5.25 Crack in a mixed mode stress field

To do that, we first define the stresses by combining Equations (5.5) and (5.11) as follows

$$\sigma_x = \frac{K_{\mathrm{I}}}{\sqrt{2\pi r}}\cos\frac{\theta}{2}\left(1-\sin\frac{\theta}{2}\sin\frac{3\theta}{2}\right) - \frac{K_{\mathrm{II}}}{\sqrt{2\pi r}}\sin\frac{\theta}{2}\left(2+\cos\frac{\theta}{2}\cos\frac{3\theta}{2}\right) \qquad [5.107]$$

$$\sigma_y = \frac{K_{\mathrm{I}}}{\sqrt{2\pi r}}\cos\frac{\theta}{2}\left(1+\sin\frac{\theta}{2}\sin\frac{3\theta}{2}\right) + \frac{K_{\mathrm{II}}}{\sqrt{2\pi r}}\sin\frac{\theta}{2}\cos\frac{\theta}{2}\cos\frac{3\theta}{2}$$

$$\tau_{xy} = \frac{K_{\mathrm{I}}}{\sqrt{2\pi r}}\sin\frac{\theta}{2}\cos\frac{\theta}{2}\cos\frac{3\theta}{2} + \frac{K_{\mathrm{II}}}{\sqrt{2\pi r}}\cos\frac{\theta}{2}\left(1-\sin\frac{\theta}{2}\sin\frac{3\theta}{2}\right)$$

For plane strain problems we have $\sigma_z = \nu(\sigma_x + \sigma_y)$ and Equations (5.107) are supplemented by the equation

$$\sigma_z = 2\nu\frac{K_{\mathrm{I}}}{\sqrt{2\pi r}}\cos\frac{\theta}{2} - 2\nu\frac{K_{\mathrm{II}}}{\sqrt{2\pi r}}\sin\frac{\theta}{2} \qquad [5.108]$$

For plane stress problems, the stresses in z-direction equal zero, that is, $\sigma_z = \tau_{zx} = \tau_{yz} = 0$.
    Let us consider the strain energy density in combined loading. The general form of the strain energy density in three dimensions (see Chapter 2) is

$$\frac{dW}{dV} = \frac{1}{2E}\left(\sigma_x^2 + \sigma_y^2 + \sigma_z^2\right) - \frac{\nu}{E}\left(\sigma_x\sigma_y + \sigma_y\sigma_z + \sigma_z\sigma_x\right) + \frac{1+\nu}{E}\left(\tau_x^2 + \tau_y^2 + \tau_z^2\right) \qquad [5.109]$$

In two-dimensional problems this equation becomes

$$\frac{dW}{dV} = \frac{1+\nu}{2E}\left[\frac{\kappa+1}{4}\left(\sigma_x + \sigma_y\right)^2 - 2\left(\sigma_x\sigma_y - \tau_{xy}^2\right)\right] \qquad [5.110]$$

Note that for plane strain κ is defined by Equation (5.9), and for plane stress by Equation (5.10). Upon introducing the stresses from Equation (5.107) into Equation (5.110), we obtain

$$\frac{dW}{dV} = \frac{1}{r}\left(a_{11}K_I^2 + 2a_{12}K_IK_{II} + a_{22}K_{II}^2\right) \qquad [5.111]$$

where factors a_{11}, a_{12} and a_{22} equal

$$a_{11} = \frac{1+v}{8E}\left[(1+\cos\theta)(\kappa-\cos\theta)\right] \qquad [5.112]$$

$$a_{12} = \frac{1+v}{8E}\sin\theta(2\cos\theta-\kappa+1)$$

$$a_{22} = \frac{1+v}{8E}\left[(\kappa+1)(1-\cos\theta)+(1+\cos\theta)(3\cos\theta-1)\right]$$

Now consider the direction of crack propagation. The derivation is based on the hypotheses presented by Sih[23] as follows:

(a) the crack grows in the direction of minimum strain energy density dW/dV;

(b) the crack propagates when dW/dV reaches a critical value $(dW/dV)_{cr}$.

The direction θ toward the minimum of strain energy density is defined by

$$\frac{d\left(\frac{dW}{dV}\right)}{d\theta} = 0, \qquad \frac{d^2\left(\frac{dW}{dV}\right)}{d\theta^2} > 0 \qquad [5.113]$$

Let us express the strain energy density, Equation (5.111), in the form

$$\frac{dW}{dV} = \frac{S(\theta)}{r} \qquad [5.114]$$

where S is a strain energy density factor defined by the equation

$$S(\theta) = a_{11}K_I^2 + 2a_{12}K_IK_{II} + a_{22}K_{II}^2 \qquad [5.115]$$

Now the direction of crack propagation θ can be obtained solving the factor S for minimum as follows

$$\frac{\partial S}{\partial \theta} = 0, \qquad \frac{\partial^2 S}{\partial \theta^2} > 0 \qquad [5.116]$$

Figure 5.26 shows a typical numerical solution for a steel plate, in the form of a function $\theta = f\,(K_{II}/K_I)$.

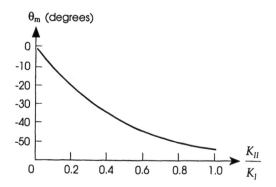

Figure 5.26 Angle of crack growth versus ratio of stress intensity factors.

The third hypothesis of Sih (Reference 23) states: (c) length r, the distance to the new site of failure, grows proportionally to S so that at any instance S/r remains constant along the new fracture surface. In mathematical form this becomes

$$\left(\frac{dW}{dV}\right)_{cr} = \frac{S_1}{r_1} = \frac{S_2}{r_2} = \ldots \frac{S_i}{r_i} = \frac{S_c}{r_c} = \text{const} \qquad [5.117]$$

Let us derive an equivalent stress intensity factor, $K_{I,eq}$, representing mixed Mode I and II. One can compare the energy density in mixed mode to the energy density of a plate loaded under uniform axial stress (Mode I). Value S is expressed by the equation

$$S = a_{11}K_{I,eq}^2 = \frac{(\kappa-1)(1+\nu)}{4E}K_{I,eq}^2 \qquad [5.118]$$

where $\theta = 0$ in the direction of crack propagation in mode I. Upon comparison one obtains the equivalent stress intensity factor becomes

$$K_{I,eq} = \sqrt{\frac{4E\left(a_{11}K_I^2 + a_{12}K_IK_{II} + a_{22}K_{II}^2\right)}{(\kappa-1)(1+\nu)}} \qquad [5.119]$$

From the latter equation one can obtain the condition at the final fracture for the mixed mode

$$\frac{4E\left(a_{11}K_I^2 + a_{12}K_IK_{II} + a_{22}K_{II}^2\right)}{(\kappa-1)(1+\nu)} = K_{Ic}^2 \qquad [5.120]$$

Figure 5.27 presents Equation (5.120) taking into consideration crack propagation direction toward the minimum of strain energy density.

Fatigue crack propagation In mixed loading the crack propagates along a path continuously changing direction. The fatigue analysis therefore must be performed in discrete steps, each one at a different angle. Because of the mixed nature, we shall sub-

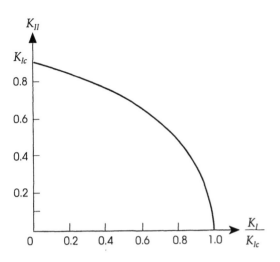

Figure 5.27 Fracture criteria in mixed modes.

stitute the stress intensity factor K for the strain energy density factor S as the basis for analysis. Consequently the fatigue crack propagation law, Equation (5.85), is replaced by the expression of Sih[24]

$$\frac{\Delta a}{\Delta N} = C(\Delta S)^n \qquad [5.121]$$

where ΔS is the change of the strain energy density factor for a given increment of crack growth, defined by

$$\Delta S = S_{i+k} - S_i \qquad [5.122]$$

Attention is directed to the fact that empirical constants C and n in Equation (5.121) are not the same as those used earlier in Equation (5.85). The solution proceeds in steps, using the numerical method described before and revised as follows:

1. Assume an initial crack size a_i and an increment of crack growth Δa_i.
2. Determine average ΔS_i.
3. Compute ΔN_i using Equation (5.121).
4. Compute next step $a_{i+1} = a_i + \Delta a_i$ and repeat steps 1 to 3 until S_c is reached.
5. Compute $N = \Sigma(\Delta N_i)$.

Note: Applying the above fatigue analysis in machine design, we assume as before (in Chapter 3) that all loading components fluctuate in phase, originating from the same source.

5.3.2 Three-dimensional cracks under one-dimensional loading

Often the cracks occur in three-dimensional bodies for which the above analysis based on two-dimensional theory is not applicable. The examples below are presented in an attempt to bring an understanding of how defining the appropriate stress intensity factors can be extended from plane solutions to fit the problems of three-dimensional bodies. These specific cases consider all the one-dimensional loading.

Elliptical crack in an infinite body   Let us refer to an embedded crack in an infinite body subjected to tension as in Figure 5.28. The stress intensity factor is given by the equation

$$K_I = \frac{\sigma \sqrt{\pi a}}{\Phi} \left(\sin^2 \varphi + \frac{a^2}{c^2} \cos^2 \varphi \right)^{\frac{1}{4}} \tag{5.123}$$

Angle φ refers to a cylindrical coordinate system concentric with the ellipse and

$$\Phi = \int_0^{\pi/2} \left(1 - \frac{c^2 - a^2}{c^2} \sin^2 \varphi \right) d\varphi \tag{5.124}$$

In a first approximation the latter becomes

$$\Phi = \frac{3}{8} + \frac{\pi}{8} \frac{a^2}{c^2} \tag{5.125}$$

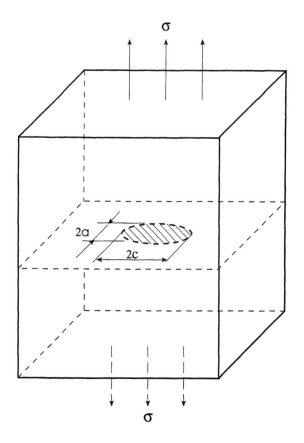

Figure 5.28 Elliptical crack in a body under one-dimensional loading.

Semi-elliptical surface crack   A body with a semi-elliptical crack is shown in Figure 5.29. The stress intensity factor is expressed by a correlation similar to Equation (5.123); however it varies along the crack front. K_I reaches the highest limit at the end of the minor axis of the ellipse, as expressed by

$$K_{I(\varphi=0)} = 1.12 \frac{\sigma\sqrt{\pi a}}{\Phi} \qquad [5.126]$$

while at the end of the major axis of the ellipse it is at the lowest,

$$K_{I(\varphi=0)} = 1.12 \frac{\sigma\sqrt{\dfrac{\pi a^2}{c}}}{\Phi} \qquad [5.127]$$

Cracks in pressure vessels   Presented here is a cylindrical pressure vessel with an axial crack, see Figure 5.30. Since the thickness of the walls is an influential factor, let us first consider a thin walled vessel. The stress intensity factor here is

$$K_I = M_F \sigma_H \sqrt{\pi a} \qquad [5.128]$$

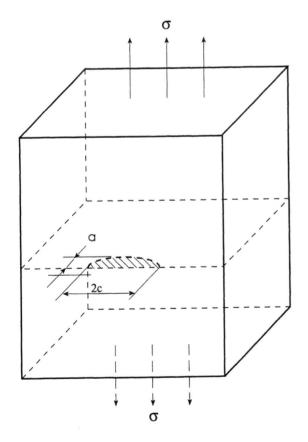

Figure 5.29 Semi-elliptical crack in a surface.

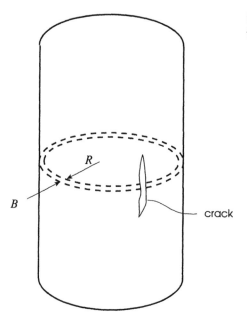

Figure 5.30 Crack in a thin-walled pressure vessel.

where σ_H is the hub stress and is expressed in the form of

$$\sigma_H = \frac{pR}{B} \qquad\qquad [5.129]$$

p is the inside pressure and M_F is a magnification factor defined by

$$M_F = \sqrt{1 + 1.61 \frac{a^2}{RB}} \qquad\qquad [5.130]$$

In a thick walled vessel we may consider the crack to be semi-elliptic. The stress intensity factor at the minor ellipse axis will be

$$K_I = 1.12 \frac{(\sigma + p)\sqrt{\pi a}}{\Phi} \qquad\qquad [5.131]$$

Additional cases with detailed correlations can be found in Broek (Reference 22).

REFERENCES

1. Irwin, G.R. 1957. "Analysis of Stresses and Strains Near the End of a Crack Traversing a Plate," *Trans. ASME, J. Appl. Mechanics*, 24.
2. Sih, G.C. 1973. *Handbook of Stress Intensity Factors for Researchers and Engineers*, Institute of Fracture and Solid Mechanics, Bethlehem, PA: Lehigh University.
3. Smith, R.A., and K.J. Miller. 1977. "Fatigue Cracks in Notches," *Int. J. Mech. Sci.*, 19, 11–22.

4. Dowling, N.E. 1979. "Fatigue at Notches and the Local Strain and Fracture Mechanics Approach," *Fracture Mechanics,* C.W. Smith (ed.), ASTM STP 677, 247–273.
5. Griffith, A.A. 1921. "The Phenomena of Rupture and Fracture in Solids," *Phil. Trans. Roy. Soc.,* London, A 221, 163–197.
6. Rice, J.R., and D.C. Drucker. 1967. "Energy Changes in Stressed Bodies Due to Crack Growth," *Int. J. Fract. Mech.,* 3, 19–27.
7. Orowan, E. 1955. "Energy Criteria of Fracture," *Welding Journal,* 34, 1575–1605.
8. Rice, J.R. 1968. "A Path Independent Integral and the Approximate Analysis of Strain Concentration by Notches and Cracks," *J. Appl. Mech.,* 35, 379.
9. Begley, J.A., and J.E. Landes. 1972. "The J Integral as a Fracture Criterion," *Fracture Toughness,* ASTM STP 514, 1–20.
10. Irwin, G.R. 1960. "Plastic Zone Near a Crack and Fracture Toughness," *Proc. 7th Sagamore Conf.,* N-63.
11. Brown, W.F., and J.E. Srawley. 1966. "Plane Strain and Crack Toughness Testing of High Strength Metallic Materials," ASTM STP 410.
12. Irwin, G.R. 1960. ASTM Bulletin, 29.
13. *Methods for Crack Opening Displacement Testing (COD).* 1972. British Standard Institution Draft for Development 19, London.
14. Hudson, C.M. 1969. NASA Tech. Note D-5390.
15. Paris, P.G., M.P. Gomez and W.E. Anderson. 1961. "A Rational Analytic Theory of Fatigue," *The Trend in Engineering,* 13, 9–14.
16. Paris, P.G., and F. Endoyan. 1963. "A Critical Approach to Crack Propagation Laws," *Trans. ASME, J. Basic Eng.,* 85, 528.
17. Barsom, J.M. 1971. "Fatigue Crack Propagation in Steels of Various Yield Strengths," *Trans. ASME, J. Eng. Ind.,* 1190.
18. Forman, P.G., V.E. Kearney and R.M. Engle. 1967. "Numerical Analysis of Crack Propagation in Cyclic-Loaded Structures," *Trans. ASME J. Basic. Eng.,* 89, 459.
19. Leis, B., D. Nelson and D. Socie. 1988. "Fatigue life Prediction," *Fatigue Design Handbook,* ed. R.C. Rice et al., Warrendale, PA: Society of Automotive Engineers.
20. Barsom, J.M. 1973. "Fatigue Crack Growth under Variable Amplitude Loading in ASTM A514 grade B Steel," ASTM STP 536.
21. Wheeler, O.E. 1972. "Spectrum Loading and Crack Growth," *Trans. ASME, J. Basic Eng.,* 94, 181.
22. Broek, D.A. 1986. *Elementary Engineering Fracture Mechanics.* Dordrecht, The Netherlands: Martinus Nijhoff.
23. Sih, G.C. 1974. "Strain Energy Density Factor Applied to Mixed Mode Crack Problem," *Int. J. Fract.,* 10, 305–322.
24. Sih, G.C. 1984. *Fracture Mechanics Methodology.* Dordrecht, The Netherlands: Martinus Nijhoff.

6

SURFACE INTEGRITY AND FATIGUE

The fatigue of a machine part depends strongly on its surface layer condition. The experience shows that fatigue failure begins in most cases at the surface. This is due to the fact that surface layers bear the greatest load and that they are exposed to environmental effects. The factors acting upon the surface layers, all causing crack initiation, are: stress concentration, oxidation, and burning out of alloy elements (at high operational temperatures). Crack initiation and propagation, in most cases, can be attributed to surface integrity which includes surface roughness, structure and stress conditions of the surface layer and material properties.

The importance of surface integrity rises as the conditions of loading, temperature, speed and vibration increase. This is true especially for high strength steels and alloys, which are more sensitive to stress concentration and surface defects. The quest for surface integrity follows from this.

6.1
SURFACE ROUGHNESS

The surface layer is conditioned by manufacturing processes, particularly, by finishing treatments. Among these processes are: chip removal processes, abrasive operations, surface deformation methods, electro-physical and electro-chemical methods, ultrasonic and laser methods. During the manufacturing process the surface layer is subjected to elastic-plastic deformation and heating, which result in structural changes, strain hardening and residual stresses, while irregularities may appear, creating surface roughness. Fatigue strength of machine parts may be improved by the choice of methods and conditions which will offer the optimal combination of surface integrity parameters.

6.1.1 Surface roughness form and measurement

The actual machine part surface, formed by a manufacturing process, is not the nominal surface as seen on the engineering drawings. The real surface always contains irregularities in the form of peaks and valleys. The multitude of irregularities, spaced relatively close to each other, represent surface roughness. The profile of the tool leaves the geometrical trace in the roughness of the surface after the mechanical processing which distorts it through plastic and elastic deformation, vibration and other factors connected with a particular process. For example, the cutting tool, while in motion, destroys the metal grains (some are cut, others are broken), creating very small cracks along the path of the tool.

Surface roughness parameters

The surface roughness is commonly determined by its profile in a cross section perpendicular to the direction of irregularities. The parameters of roughness are evaluated on the basis of a part of a profile, called the sampling length. The most important parameters describing surface roughness, specified by international standards, are:

R_a = arithmetic average roughness height,
R_q = root mean square (RMS) roughness height,
R_z = ten-point height (average distance between five highest peaks and five deepest valleys),
R_{max}, R_y = maximum peak-to-valley roughness height,
R_p = maximum peak height,
A_r = average spacing of roughness peaks.

Since the valleys between irregularities of a surface are essentially notches that weaken the surface and cause stress concentrations, other important parameters from the fatigue point of view are mean valley radius r_v and maximum valley depth R_v. See Figure 6.1. A smaller radius r_v, indicates a sharper notch, while a greater R_v represents a deeper valley—both disclose higher stress concentration.

Unfortunately, parameters r_v and R_v are used mainly by researchers and are not included in standards. In practice, the most frequently used parameters are R_a, R_z and R_{max} which describe the height of irregularities and give only an indirect indication of the sharpness and depth of surface notches. However, they provide enough insight to predict fatigue behavior of the designed part. The ratio of R_{max} to R_a is about[1]

$$\frac{R_{max}}{R_a} \cong 4 \qquad\qquad [6.1]$$

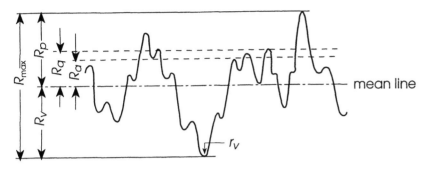

Figure 6.1 Parameters describing surface roughness.

Surface roughness measurement

Surface roughness evaluation may be performed by both quantitative and qualitative methods. Qualitative methods are based on comparing the processed surface with roughness specimen made of similar material and processed by similar methods. The inspection is visual or manual, by moving a finger nail transverse to the processing traces. This method reveals surface roughness up to $R_a = 0.63$ μm. On smoother surfaces a comparison is made with roughness specimen using microscope-type comparators. This type of test is simple and is applied widely in practice. Quantitative methods of evaluating surface roughness pertain to measurements of micro-irregularities by special instruments and include both, contacting and noncontacting measuring methods.

The main contacting method is by tracing the inspected surface with a diamond stylus. Instruments based on this method include profilometers, which give a direct indication of the roughness parameter on a numerical scale, as well as profilographs, which record the surface profile as graphs. Profilometers measure surface roughness within the limits of $R_a = 5$ to 0.04 μm. Surface roughness of smoother surfaces, up to $R_a = 0.01$ μm, is recorded on profilographs and the numerical values of R_a are determined from the graphs.

Noncontacting measurements employ optical instruments based on the surface light section method, measuring surface roughness up to $R_a = 0.63$ μm, or using instruments based on the light interference method (microinterferometers) capable of measuring irregularities up to $R_a = 0.03$ μm.

In cases where the measured area is inaccessible and instruments can not be used for measurement of surface roughness, for example, large-sized parts and inner surfaces in deep apertures, the replica method is employed. The measured surface is covered with a special plastic mass. After it congeals, the mass is removed and used as facsimile reflecting the irregularities. By measuring the surface roughness of the replica, one obtains surface roughness parameters of the examined part.

6.1.2 Effect of processing on surface roughness

The surface roughness of a machine part is shaped during its manufacture, mainly during finish, being influenced by the methods and conditions of the processing. Figure 6.2 illustrates the different shapes of irregularities resulting from different processing

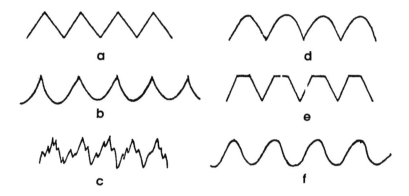

Figure 6.2 Models of surface irregularities processed by different methods: (a) cutting with sharp tool; (b) cutting with round-nose tool; (c) grinding; (d) polishing; (e) superfinish; (f) rolling and burnishing.

conditions and methods. Obviously, surfaces with different irregularities will have different effects on fatigue properties.

Known methods of producing machine part surfaces include: chip removal operations (turning, milling, boring, planing, and others), abrasive processes (grinding, polishing, honing, and superfinish) and finishing processes (rolling, shot-peening, and burnishing). Let us review the effects of the different finish methods on the surface roughness.

Chip removal operation

Factors influencing surface roughness of a machine part produced by a cutting tool may be grouped into three basic categories:

(a) the geometry of the cutting tool and its path,
(b) elastic and plastic deformation of the processed material,
(c) vibration of cutting tool.

The size and shape of irregularities are geometrically determined by the shape and path of the cutting edge which are duplicated in the processed surface. The plastic and elastic deformations in the processed material, as well as the cutting tool's vibrations distort the geometrically regular pattern and may considerably increase the height of irregularities.

Geometrical causes of irregularities   During processing, at each revolution of the processed part (in turning or boring.) or tool (in milling), the cutting edge cuts a groove and is shifted a distance of feed f from position 1 to position 2, leaving a wave pattern on the surface of the processed part. See Figure 6.3(a). The size and shape of such irregularities depend on feed and the shape of cutting edge. When the feed is shortened the height of irregularities decreases, Figure 6.3(b), and when the feed is lengthened their height increases. A change in cutting tool angles φ and φ_1 increases or decreases both the height of irregularities and their shape. See Figure 6.3(c). Use of a cutting tool with rounded nose produces lower irregularities comprising rounded valleys. See Figure

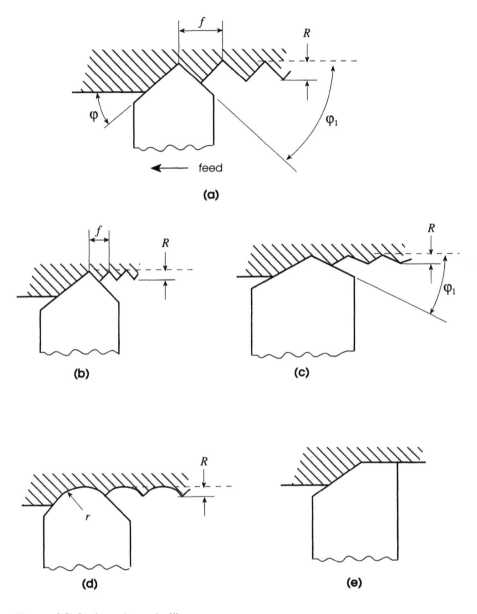

Figure 6.3 Surface irregularities.

6.3(d). A cutting edge, parallel to the direction of feed, theoretically produces a smooth wave-free surface. See Figure 6.3(e).

The following expressions, based on geometrical analysis, can be used to estimate the height of irregularities. For sharp cutting tools without curvature the arithmetic average height equals

$$R_a = \frac{\tan \varphi \cdot \tan \varphi_1}{4(\tan \varphi \cdot \tan \varphi_1)} \cdot f \qquad [6.2]$$

For cutting tools with a rounded nose with radius $r > f$ the height is

$$R_a = \frac{f^2}{32r}$$ [6.3]

For cutting tools with a rounded nose with radius $r < f$ the height is

$$R_a = \frac{\sin \varphi \cdot \sin \varphi_1}{4\sin\left(\varphi + \varphi_1\right)}\left[f - r\left(\tan \frac{\varphi}{2} + \tan \frac{\varphi_1}{2}\right)\right]$$ [6.4]

Geometrical factors affecting surface roughness also include tool chipping, causing jags in the cutting edge which are reproduced on the processed surface. The greatest effect is in precision processing, when the size of jags is comparable to the size of required surface roughness. The dullness of the tool and chipping increase surface roughness by about 20 to 60 percent, depending on processing conditions (type of tool, processed material, processing condition and duration). To reduce the chipping effect of the tool, its timely replacement is necessary.

It follows from geometrical analysis that, to reduce irregularities, feed reduction, small φ and φ_1 angles, and a rounded nose are required. A proper choice of the cutting tool geometry provides control of the surface roughness over a wide range of limits. Yet, even with minimal feed and optimal tool geometry, one cannot eliminate entirely the surface roughness, because of the effect of plastic deformation of the processed material.

Irregularities caused by plastic deformation   During the chip removal process, along with the presence of chips from cut-off material, a plastic deformation of the surface layer takes place. As a result of tool pressure, the material undergoes plastic deformation distorting the surface irregularities determined by the geometry of the process and increasing the height of irregularities. See Figure 6.4.

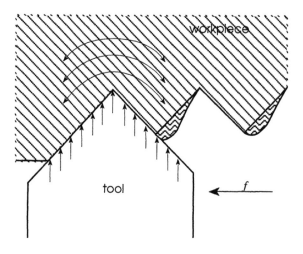

Figure 6.4 Plastic deformation of geometric irregularities.

The extent of plastic distortion of irregularities is affected strongly by the type of processed material and its structure. Cutting materials of greater ductility and a heterogeneous structure with large grains will cause coarser and rougher surfaces. Heat treatment (such as normalizing, hardening and tempering) will create a homogeneous small-grain structure while increasing its hardness and, after subsequent mechanical processing, will produce much lower roughness.

A dull cutting blade requires that an increased cutting force and pressure be applied to the processed surface. This, in turn, increases the magnitude of plastic deformation and the height of the irregularities.

In most metals with greater ductility and at certain temperatures (300 to 500°C for carbon and low-alloy steels), a built-up edge of welded particles is formed on the front face of the cutting tool. This built-up edge distorts the geometry of the blade, impairing the surface condition. Blades with a built-up edge cause greater roughness of the processed surface. See Figure 6.5. To eliminate the built-up edge, cutting speeds must be changed so that the temperature will be out of the danger zone. Built-up edges usually disappear at speeds exceeding 60–70 m/min or under 2–5 m/min. Another way to eliminate the built-up edge and reduce surface roughness is by applying lubricants and coolants with active additives (such as sulfur and oleine acid). These liquids effectively keep the surfaces of the cutting tool and the machined part separated, preventing sticking and formation of a built-up edge.

Vibration of the cutting tool During the chip removal process vibrations take place within a system comprised of the processed part, the cutting tool and the tool operating machine. The vibrations are caused by oscillating external forces (for example, spindle imbalance) and the nonuniformity of the cutting process (that is, self-sustained vibrations). As a result of the process kinematics, vibration of the cutting tool in relation to the processed surface distorts the geometrical pattern of irregularities. The larger the amplitude of vibrations, the greater the height of surface roughness. The most pronounced effect on the surface roughness from system vibrations occurs in precision processing, when other factors are minimal.

The way to reduce the effect of vibrations on surface roughness is to reduce the amplitude of vibrations of the system. This requires optimal use of machinery and fixtures with high rigidity and vibration resistance, isolated from other sources of vibration, and with special vibration damping devices applied.

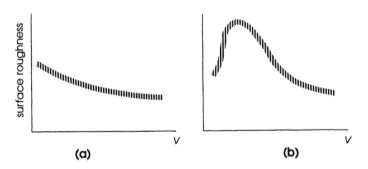

Figure 6.5 Surface roughness versus velocity *v* in processing of materials: (a) not inclined built-up edge; (b) inclined built -up edge.

Abrasive processes

Abrasive methods are the more popular among surface finish processes. During abrasive processing, the cutting operation is performed by an undefined large number of abrasive grains having different shapes and scattered randomly with respect to each other. The angles between faces of abrasive grains range from 40 to 150 degrees. The most prevalent abrasive processing methods include grinding, honing, polishing, lapping and superfinish.

The abrasive tool is composed of particles with different degrees of sharpness and protrusion. The sharpest and most protruding particles scrape the processed surface, forming a large number of deep grooves. Other duller grains of the abrasive tool, protruding less, do not scrape the processed surface rather plastically deform it instead and press out a large number of valleys. The rest of the grains, dull and protruding even less from the tool surface, interact with the processed surface in an elastic friction mode. This type of interaction produces an irregular surface texture with a small height of irregularities (R_a, R_z), small peak spacing (A_r), and large individual peaks and valleys (R_{max}, R_v). The bottom of valleys, scraped out by abrasive grains, is sharper than those created in a chip removal process, having a smaller round-off radius r_v. Figure 6.6 shows typical profiles of surfaces processed by abrasive applications.

The surface irregularities are determined by the geometry of abrasive grains and by the kinematics of the tool and the processed part. The geometrical parameters affecting surface roughness in abrasive processing include the size of abrasive grains, the

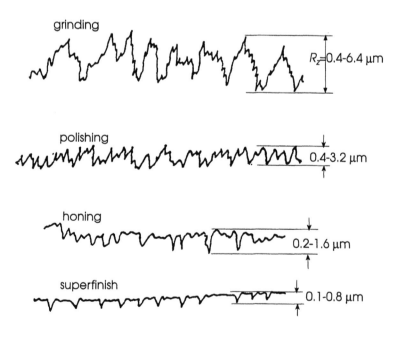

Figure 6.6 Profiles of surfaces processed by abrasive operations.

rotation speed of the abrasive tool, and the kinematics of its motion relative to the processed surface. The larger the grains, the deeper and wider are the created grooves, resulting in higher and wider irregularities. As the size of grains in the tool varies, the larger grains create the deepest grooves. When the rotational speed of the abrasive tool is increased and the linear speed relative to the processed part is reduced, the number of scratches per surface unit is increased and the height of irregularities reduced.

Determined by the geometry of cutting grains and the relative speeds between the tool and workpiece, the surface irregularities are distorted by the action of contact pressure and heating in the processing area. Processing softer and more ductile materials results in greater distortions.

The abrasive process, particularly grinding and polishing, is accompanied by vibrations. The vibrations, in general, are caused by an unbalanced grinding wheel. As the imbalance increases, the level of vibrations rises, resulting in larger irregularities on the processed surface.

The surface roughness obtained in grinding is in the range of $R_a = 0.2$–$1.0\ \mu m$, depending on the processed material and processing modes. A lower surface roughness is achieved when using precision abrasive methods such as honing, polishing, lapping, and superfinish. By using small abrasive grains that result in weaker forces and smaller indentations in the processed surface, a surface roughness in the range of $R_a = 0.02$ to $0.16\ \mu m$ can be achieved.

Depending on processing methods, the initial surface is, in general, either totally removed and a new surface roughness is formed comprising small scratches with narrow spacing, or the initial surface is removed only partially and the final roughness is a combination of old and new irregularities. See Figure 6.6.

Cold-working processes

Cold-working processes (also called strain-hardening processes) involve plastic deformation of surface irregularities without cutting. Cold working is performed either by using a pressing tool (rolling, burnishing, mandrelling) or by impact onto the processed surface (shot peening, centrifugating). See Figure 6.7. Since the cold-working

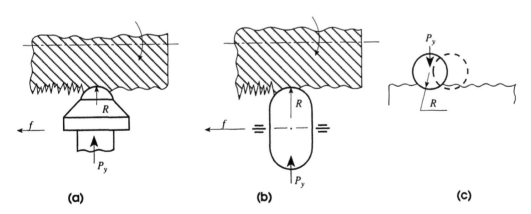

Figure 6.7 Creation of surface irregularities in cold-working operations: (a) burnishing; (b) rolling; (c) shot peening.

methods do not need a cutting edge, the tools used here have a round operating face (such as rollers, burnishers, shot, and balls).

The effect of a cold-working process on surface roughness depends on the geometrical shape of the operating face of the tool and the kinematics of its motion relative to the processed surface. In rolling and burnishing (as in turning), a spiral groove is formed on the processed surface and surface irregularities are formed as peaks and valleys. In shot peening, shots striking the surface create a multitude of craters with crests round their edges. An additional factor affecting the surface roughness in cold-working methods is the plastic flow of surface material under a static or impact pressure. The displaced material from the pressure zone forms new surface irregularities.

The size of irregularities in processing by the cold-working method depends on the size of the initial surface roughness and the type of processing method applied. For example, in shot peening the range of a surface roughness that can be achieved is $R_a = 0.63-5.0 \, \mu$m, while in burnishing it can be $R_a = 0.02-0.3 \, \mu$m.

The texture of surface processed by cold working may be characterized as follows: first, unusually inclined and round shaped irregularities (as compared with cutting); second, large valley radiuses r_v; and third, widely spaced irregularities, relative to their height. The latter is caused by operating faces with large radiuses (balls, rollers, burnishers, shots) because the tool radiuses used in cold working are much larger than those used in chip removal or abrasive processing. The surface roughness features obtained in cold-working processing increase fatigue strength of the machine part, creating a favorable condition for operations under alternating loads.

6.1.3 Change of surface roughness in service

The surface roughness of an operating machine part changes while in service. Therefore, evaluation of the fatigue strength must be based on a surface roughness at a later stage, not based on the initial surface roughness obtained in manufacturing. The changes in surface roughness are affected by specific conditions under which a machine part operates—pressure, speed, friction and others.

Changes of surface roughness due to friction are common in sliding machine parts. The biggest change occurs at the initial stage—the run-in period. During this period the existing irregularities are substantially worn down and a new surface roughness is formed, optimal for the friction condition. After the run-in period (in spite of the persisting wearing off conditions) the surface roughness remains practically unchanged. The new roughness of the surface after run-in is independent of the initial surface roughness. In cases where the initial surface roughness is similar to the new one, the duration of the run-in period is minimal. Optimal parameters for the initial surface roughness are determined experimentally. Table 6.1 lists optimal surface roughness parameters R_a, for several sliding machine parts.

Surface roughness in fixed joints changes due to fretting corrosion. It occurs at high contact pressures and extremely small cyclic displacements of surfaces in contact. Such conditions are typical for smooth and spline joints under cyclic loading. In the condition of fretting corrosion, the initial surface deteriorates rapidly—numerous depressions (craters) are formed on the surface, reducing significantly the fatigue strength of the involved machine parts.

TABLE 6.1
Optimal Surface Roughness for Sliding Machine Parts

Machine Part	R_a (μm)
Turbine shaft in gas-turbine engine	0.3–0.63
Compressor rotor in gas-turbine engine	0.3–0.63
Crankshaft—automotive	0.3–0.63
Camshaft—automotive	0.3–0.63
Universal joint	0.2–0.8
Car axle	0.63–2.50
Turbine blade in gas-turbine engine	0.25–0.32
Compressor blade in gas turbine engine	0.5–0.63

Changes of surface roughness in parts exposed to air or gas flow are mainly due to surface erosion and corrosion. A typical example is a surface erosion in turbine blades of turbojet engines. The blade surface is exposed to a high-speed gas flow which causes erosional-corrosional damage to the surface layer. A nonuniformity of gas velocities and pressures at different surface locations causes uneven changes in the surface roughness. At high gas temperatures surface oxidation increases, erosion is accelerated, pittings and microcracks appear, and, as a result, the surface roughness deteriorates. After the run-in period, experiments with blades of different initial roughness (produced by different methods—turning, milling, grinding, polishing), have shown that a changed surface roughness of $R_a = 0.8$–$2.5\ \mu m$ is obtained independent of the initial one. A polished surface with an initial roughness of $R_a = 0.08$–$0.32\ \mu m$ changed most, reaching a surface roughness of $R_a = 0.8\ \mu m$. See Figure 6.8.

The surface roughness of machine parts changes while in service, usually deteriorating. Therefore one must consider the roughness at the operational stage and not the initial surface roughness when estimating the fatigue strength of a working machine part.

6.1.4 Surface roughness and fatigue

The processed surface of a machine part comprises a large number of grooves (notches) of different depths and sharpness—causing local stress concentrations and reducing the fatigue strength. Based on tests, it is known that the detrimental effect of stress concentration is greater in materials with a higher ultimate strength. Therefore, when made of such materials, machine parts subject to fatigue demand special consideration in mechanical processing.

The theoretical dependence between the stress concentration factor and the size of surface grooves is defined by Neuber rule. (See Chapter 3.) In application to the surface roughness, the stress concentration factor can be expressed in the empirical form[2]

$$K_t = 1 + n\sqrt{\gamma \frac{R_z}{r_v}}$$

[6.5]

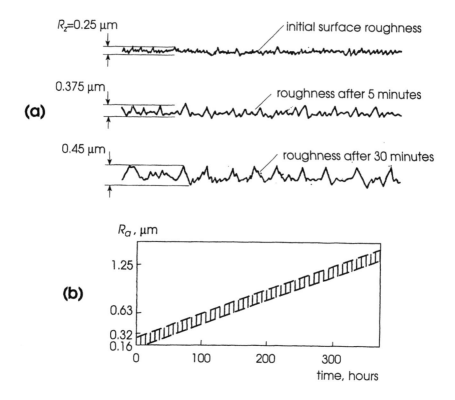

Figure 6.8 Surface roughness deterioration of turbojet engine blade: (a) deterioration during first half hour; (b) long term deterioration. (Data from Sulima, A. M., and M I Evstigneev, see Reference 3.)

Coefficient γ depends on the ratio of spacing and height of irregularities (for surfaces after mechanical processing $\gamma = 1$). Coefficient n depends on the kind of stresses: $n = 1$ in shear and $n = 2$ in tension and bending. Based on the above, the stress concentration factor for ground surfaces with $R_a = 0.32\ \mu$m equals 1.2; for ground surfaces with $R_a = 1.25\ \mu$m it equals 1.48. Surfaces processed with a less refined method have a higher stress concentration factor. The above equation is difficult to apply in practice, because the curvature radius of the valley, r_v, is rarely measured.

Tests with machine parts confirm the significant influence that surface roughness has on fatigue life. The surface roughness is, however, only one of many factors. The condition of metal structure in the surface layer, strain hardening, and residual stresses, all influence greatly the fatigue life of a part. In experimental studies different surface roughness is obtained using different processing methods which change not only the surface roughness but structural condition of material and add residual stresses in the surface layer as well.

Experiments to determine the effect of surface roughness on fatigue were conducted by Sulima[3] with specimens of high-temperature resistant Cr-Ni alloys. To obtain different roughness, the specimens were ground by an abrasive wheel, abrasive belt, rolled and milled. In order to isolate the effect of surface roughness on fatigue and to exclude effects of cold work and residual stresses, the specimens were subjected to isothermal heating in vacuum after the mechanical processing. This restored the initial undeformed structure in the surface layer and removed residual stresses generated during the mechanical processing. The tests were conducted at a temperature of 800°C. The test results are shown in Figure 6.9. The results show that the size of irregularities on the processed surface has a profound effect on fatigue—when surface roughness was increased to $R_a = 5$ μm, the fatigue limit dropped by 20 to 25 percent. The tests also showed that fatigue is affected by the direction of surface irregularities. For similar surface roughness, fatigue strength of a part with irregularities in a direction perpendicular to the axis of the part is 1.5 times lower than when the direction of irregularities is along the part's axis (see Figure 6.9). The correlation of the fatigue limit of high-temperature resistant alloys and the surface roughness can be expressed by the empirical equation

$$S_f = 25R_a^n \qquad\qquad [6.6]$$

where $n = -0.02$ for a longitudinal direction of irregularities; $n = -0.04$ for the lateral direction.

A review of test results with structural steels under normal testing conditions (temperature 20°C, loading frequency up to 100 Hz), shows that the fatigue limit drops by 10 to 25 percent when the surface roughness increases from $R_a = 0.2$ μm to $R_a = 2.5$ μm. For coarser surface roughness, the fatigue endurance drops even lower.

The fatigue strength depends not only on the size of irregularities, R_a, but also on valley radius, r_v, inclination angles of the sides of irregularities and spacing A_r. Unfortunately, these parameters are not measured in practice and no experimental data is available. Yet it is evident that larger radii of valleys and more inclined sides of irregu-

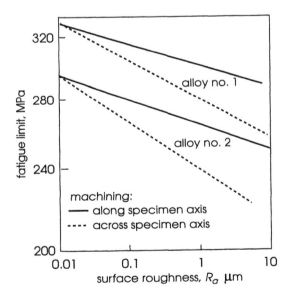

Figure 6.9 Fatigue limit versus surface roughness of a high-temperature resistant Cr-Ni alloy tested at 800°C. (Data from Sulima, A. M., and M I Evstigneev, see Reference 3.)

larities cause lesser stress concentrations, resulting in higher fatigue limit. To increase fatigue limits of machine parts one must reduce the height of irregularities (R_a, R_z, and especially R_{max}, since discrete deeper valleys cause most stress concentration), increase the valley radius r_v, and ensure the direction of irregularities to be parallel to the axis of the part. The solution to this problem is achieved through the selection of an appropriate method and mode of processing.

6.2
METAL STRUCTURE AND STRAIN HARDENING OF THE SURFACE LAYER

During processing the workpiece is subjected to a wide range of processing and environmental influences that alter its surface layer properties and affect its fatigue strength. To name a few that can cause a diminished fatigue resistance, are working tool pressure, the friction between a tool and the processed surface, and the effects of coolants and gaseous environments. Different processing methods change not only the surface roughness, but structural condition of material as well and add residual stresses in the surface layer. The condition of metal structure in the surface layer, strain hardening, and residual stresses, all influence greatly the fatigue life of a part.

6.2.1 Surface layer formation

Under the action of the cutting force and friction, plastic deformation takes place in the surface layer. This deformation gradually diminishes with the depth of the surface layer. The structure of the deformed surface layer is shown in Figure 6.10. Notice a sharp growth of the number of crystal lattice imperfections such as dislocations and vacancies. The grains are crushed and their shape and size are changed, at the surface they are refined and stretched in the direction of deformation. These changes hinder further deformation of the metal.

Plastic deformation of the metal alters its strength, hardness, elasticity, ductility and creep. As the plastic deformation increases, the strength properties such as yield point,

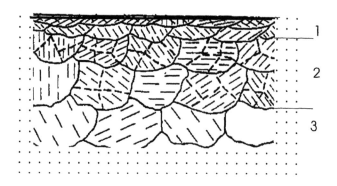

Figure 6.10 Deformation of surface layer:
(1) broken and stretched grains; (2) completely
deformed grains; (3) partially deformed grains.

fatigue strength, hardness and microhardness increase up to a certain level and the elasticity modul changes. The ductility properties such as elongation and area reduction decrease. As a result of plastic deformation the metal is strain hardened, meaning its resistance to deformation increases. In processing quench-hardened steels with a martensite-austenite structure, an additional strengthening occurs which is associated with decomposition of residual austenite and its transformation into martensite, thus forming highly disperse carbides which block slip-plane shifts. When the pressure in the contact zone between tool and workpiece rises to a certain level, the lattice distortions reach a limiting value and the material deteriorates. Dislocation and vacancy buildups grow into microcracks, causing loss of strength of the material and its failure.

Along with surface layer deformation caused by cutting and friction forces, the processing causes heat release. The heat generated by plastic deformation and external friction in the contact zone of the tool and processed material, significantly affects the surface layer and its properties. As metal temperature rises, the oscillation frequency and energy as well as atom mobility increase and the diffusion becomes more intensive. As a result, the dislocation density is reduced and plastic deformation set free. Thus, while the increased plasticity promotes a more intensive strain hardening, it also causes softening due to recovery and recrystallization (the metal returns to a minimal free-energy condition). Under normal conditions the surface layer recovers slowly, but this process accelerates as the temperature increases. For structural steels and titanium alloys it happens when the temperature reaches 400 to 550°C. For high-temperature resistant alloys it happens when the temperature reaches 700 to 900°C. At the recrystallization temperature, strain hardening is removed completely and the initial properties of the metal are restored or a new coarse grain structure is formed with weaker boundaries and reduced hardness. The recrystallization temperature of the metal depends on the melting point temperature and can be represented by the following expression:

$$T_r \cong 0.4 T_m \qquad\qquad [6.7]$$

where T_r and T_m are the recrystallization and melting point temperatures in degrees Kelvin. Thus, for example, the recrystallization temperature for lead is under 0°C; for aluminum, copper, iron, and nickel-base alloys about 100°C, 270°C, 450°C and 750°C, respectively.

Two mutually exclusive processes therefore occur within the surface layer of a workpiece during processing: strain hardening caused by plastic deformation and softening caused by heating. The final state of the surface layer state is determined by conditions and modes of processing.

6.2.2 Strain-hardening parameters

Strain hardening of the surface layer is defined by strain-hardening depth h_s, strain-hardening rate U_s and strain-hardening gradient G_s. These parameters are determined either by measurements of microhardness or by X-ray analysis methods.

Microhardness H_μ is measured using a microhardness meter on taper sections of the measured surface. See Figure 6.11. An inclined surface at an angle of $\varepsilon = 1.5 - 3$ degrees is produced by lapping and subsequent polishing with a cloth soaked in aluminum oxide.

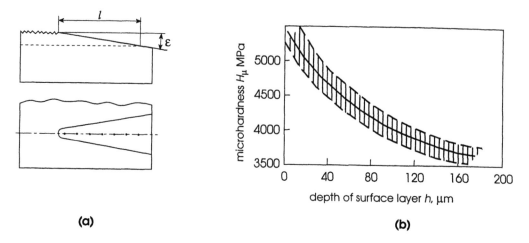

(a)   (b)

Figure 6.11 Measurement of microhardness in surface layer by taper sectioning: (a) points of measurement in taper section; (b) microhardness distribution. $H_{\mu max}$=5500MPa, h_s=128µm, U_s=44.8%.

The microhardness is measured by the measuring instrument as follows. A tetrahedral diamond pyramid with an angle of 136 degrees is pressed into the surface with a force of $P = 0.5-2$ N. Microhardness H_μ is determined by the mean value of diagonals. The measured microhardness is plotted as a function of surface layer depth. See Figure 6.11(b). The diagram is used to compute the three parameters:

$$h_s = l \tan \varepsilon \tag{6.8}$$

$$U_s = \frac{H_{\mu max} - H_{\mu min}}{H_{\mu min}} 100\% \tag{6.9}$$

$$G_s = \frac{H_{\mu max} - H_{\mu min}}{h_s} 100\% \tag{6.10}$$

Another similar technique pertains to layer-by-layer electrolytic etching with successive measurements of microhardness.

X-ray analysis method is based on measuring the width of the diffractional line of the surface. Thin layers are successively removed from the measured surface by an electrolytic technique and each time the diffractional line and its width are registered. The depth of the layer is computed with high accuracy from variations in the width of the diffraction line.

Strain-hardening parameters in mechanical processing may vary over a wide range, depending on the magnitude and frequency of applied loads, the shape and dimensions of the tool and the properties of the processed material. The depth of the strain-hardened layer, h_s, may, in fact, vary within the limits of 0.01–0.5 mm. The rate of strain hardening, U_s, for hard metals ranges from 10 to 20 percent, for structural steels 20 to 40 percent, and for high-temperature resistant alloys and stainless steels 30 to 70 percent. The gradient of strain hardening, G_s, may vary over a range 15–70 kN/mm³.

6.2.3 Effect of processing on the surface layer (strain hardening)

As explained, the surface layer condition is determined by the relation between strain hardening and softening, depending on the magnitude of plastic deformation or heating present in the processed area. The relation between these two processes depends on the type and condition of the mechanical process and the geometry of the tool used. Any change in the conditions of processing which increases the cutting force and the plastic deformation raises the rate of strain hardening. In addition, a prolonged action of the cutting force causes a greater strain-hardening depth. On the contrary, a change in processing that increases the heat in the processed area or increases the duration of the thermal effect intensifies the process of softening. These are simplified guidelines for assessment of processing effects on strain hardening of the surface layer.

In reality, determining the effect of processing on the surface layer is much more complicated due to numerous factors such as friction, heat removal conditions, mechanical properties and structural condition of the material. In order to find the optimum processing for attaining the needed strain hardening, it is necessary to know the effects of the different factors and be able to recognize the predominant ones.

Structural condition of the material

One of the factors affecting the strain hardening of the material is its structural condition. The phenomena of strain hardening have been studied using different metals, such as carbon and alloy steels in different structural states, aluminum and copper alloys. The studies confirmed a correlation between strain hardening and the structural condition of the material. Figure 6.12 shows the microhardness as a function of surface layer

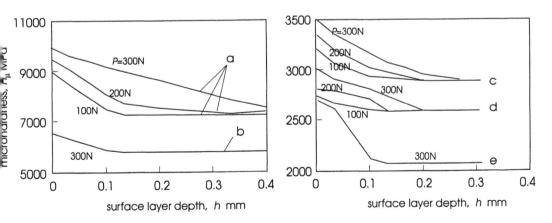

Figure 6.12 Microhardness of surface layer versus burnishing force at different structural conditions: (a) martensite, (b) troostite, (c) sorbite, (d) ferrite and perlite, (e) ferrite. Material carbon steel SAE 1045.

thickness after burnishing of carbon steels AISI 1045 in different structural states. Figure 6.13 presents the relationship among the following three factors: hardness, rate of strain hardening and structural state of the material.

Maximum rate of strain hardening takes place during processing of steels having ferrite, austenite and martensite structures. Those steels having sorbite and troostite structures have a minimum rate, explained probably by a difference in stability of these structures, where the strain hardening depends mainly on the increased density of crystal structure defects. In steels having a martensite structure (in addition to increased density of the structural defects) there is a release of dispersed carbides; an additional strain hardening takes place through transformation of residual austenite into martensite. The rate of strain hardening of a martensite structure is consequently higher. The higher the residual austenite content in the processed martensite structure steel, the higher the strain-hardening rate. Enhanced strain hardening of austenite structure steels is also explained by the austenite-martensite transformation that occurs in the surface layer under the influence of deformation. Table 6.2 shows the rate of strain hardening of steels and alloys in different structural states, achieved by mechanical processing.

Chip removal operations

The chip removal operations include turning, milling, drilling, boring, broaching, and reaming. Let us consider the effects of the most common operation, turning. As a rule, during turning, strain hardening takes place in a thin surface layer. Figure 6.11(b) shows a typical variation of microhardness in the surface layer after turning. Maximum hardness is noted at the surface, dropping gradually to its initial value. The most important parameters of a chip removal operation that affect surface layer hardness are feed and velocity. The feed exerts a force on the surface layer. The larger the feed, the higher are both the rate of strain hardening and the strain-hardening depth. See Figure 6.14(a). The cutting velocity affects the intensity and duration of heating the surface layer so that the

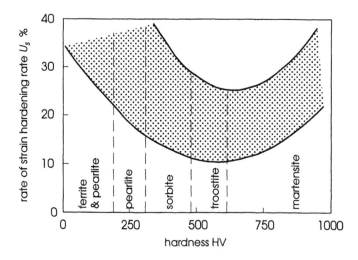

Figure 6.13 Rate of strain hardening versus hardness of carbon steel AISI 1045.

TABLE 6.2
Rate of Strain Hardening in Different Materials in
Mechanical Processing

Material	Structure	U_s
Steel	Austenite	0.50–0.60
Steel	Ferrite and Pearlite	0.35–0.45
Steel	Pearlite	0.25–0.35
Steel	Sorbite	0.20–0.30
Steel	Troostite	0.20–0.30
Steel	Martensite	0.35–0.45
Alluminum alloys		0.15–0.20
Copper alloys		0.15–0.20

higher the velocity the lower the rate of strain hardening. The depth of strain hardening decreases proportionally to heating. See Figure 6.14(b). An increased cutting depth
creates larger forces which result in a higher plastic deformation and greater surface layer
hardness. The geometry of the cutting tool also has a significant effect on surface layer
hardness. Changes in tool geometry, (a reduced cutting edge angle, an increased tool
radius, a dull cutting edge) that enlarge the cutting force, cause increased deformation
and surface layer hardness. The same principles apply to other chip removal operations.

In an abusive operation (deep cut, large feed, low cutting velocity) an intensive
plastic deformation and moderate heating take place. As a result, a higher rate of strain
hardening (up to 50 percent) and a greater strain-hardening depth (up to 0.2–0.4 mm)

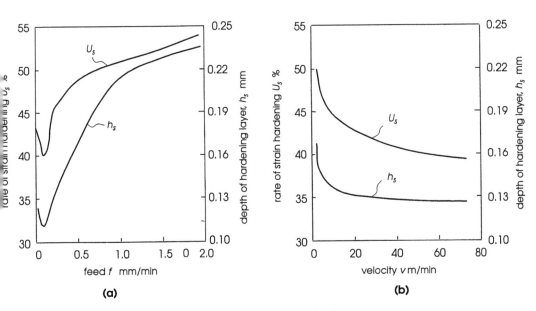

(a)   (b)

Figure 6.14 Rate of hardness and depth of strain hardening
versus tool feed and velocity in cutting of steel.

occur at the surface layer. In a gentle operation (low cutting depth, small feed, sharp cutting tool, high cutting velocity) plastic deformation is limited to a thinner surface layer and the heating significantly reduces (or almost eliminates) the strain hardening of the layer.

Abrasive machining

Abrasive machining includes grinding, honing, and polishing. Here the same principles apply: strain hardening increases proportionally to the force exerted on the abrasive grain, and is in accordance with grinding depth, feed, and geometry of the grains.

When a rotation speed of a grinding-wheel increases, the load on the grains is reduced and the released heat increases, whereby strain hardening of the surface layer diminishes or even disappears. The heating effect in abrasive machining is greater than that in a chip removal operation because of a higher cutting speed. In processing non-hardened metals and alloys, the temperature developed in the processed area does not usually cause phase transformations. In such metals the rate of strain hardening may reach 30 to 40 percent and the strain-hardening depth up to 0.02–0.04 mm, as per grinding conditions.

In grinding quench-hardened steels with martensite or troostite-martensite structure, phase transformation may occur depending on surface layer heating. Refer to Figure 6.15. At a certain combination of the applied cutting forces and temperatures (the latter not exceeding transformation temperature Ac_3 in an iron-carbon equilibrium diagram), a strain-hardened layer is formed with enhanced mechanical properties—higher hardness and strength. See curve b in Figure 6.15. In grinding at higher speeds with insufficient cooling, the large amount of generated heat (in combination with the applied forces) may cause structural phase transformations either locally or over the entire area with consequent changes in hardness and color and the formation of microcracks. The local

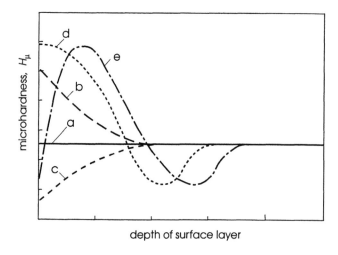

depth of surface layer

Figure 6.15 Distribution of microhardness in surface layer: (a) initial, (b) strain hardening, (c) tempering burn, (d) secondary hardening, (e) tempering and secondary hardening.

structural changes of this kind are called burns. When the temperature is increased to Ac_3 point, tempering of martensite occurs with a formation of pearlite-troostite structure, developing tempering burns; here the microhardness becomes lower than the initial value. See curve c in Figure 6.15. When the grinding temperature exceeds the Ac_3 temperature, an intensive heating takes place with subsequent rapid cooling which causes secondary strain hardening of a thin surface layer. The secondary strain-hardened layer (up to 0.01 mm thick) has a high microhardness, that is higher than the original value. Usually, a reduced microhardness layer is found under the secondary strain-hardened layer. See curve d in Figure 6.15. At a more intense heating, where the heat is not absorbed fast enough by internal layers, a tempering of the secondary strain-hardened layer occurs. The distribution of microhardness in the surface layer in this case is represented by curve e in Figure 6.15.

The burns cause stress concentrations that reduce fatigue strength. Since the high temperature is the main source of burn and crack formation in the processed surface, temperature reduction can improve surface quality. Some of the most effective means to reduce temperature include: the proper choice of grinding regimes (especially the wheel speed), wheel dressing properties, composition of a grinding fluid, and the fluid introduction method.

In abrasive finishing, such as honing, lapping, polishing and superfinish, the heating of the surface layer is lower due to the low processing speeds and the surface layer becomes strain hardened. The strain-hardening depth is limited to 0.005–0.02 mm because of small grains and forces, but the rate of strain hardening is high (reaching 15 to 40 percent for steels). In fact, in polishing done with soft felt or graphite wheels strain hardening may approach zero. In electropolishing the strain-hardening is totally eliminated.

Cold-working

Surface processing methods by cold-working (also called strain-hardening processing) include rolling, burnishing, and shot peening. The methods are characterized by high pressures in the contact area (between the tool and the workpiece) and very low heating which results in significant strain hardening of the surface layer. Figure 6.12 presents values of microhardness in surface layers of different metals obtained by burnishing. The rate of strain hardening and the thickness of strain-hardened layer depend on the properties of the processed material (its structural phase has the greatest effect). Table 6.2 shows rates of strain hardening obtained by processing of steels and alloys with different structures. The most hardened are austenite and martensite steels, the least hardened are sorbite and troostite steels and aluminum and copper alloys.

Strain hardening of the surface layer depends on the strain-hardening process. In static deformation methods, such as rolling and burnishing, the strain hardening is intensified by the following factors: increased deformation force, prolonged time and a larger number of tool passes. Figure 6.16 presents the strain-hardening parameters (H_μ, U_s, h_s) plotted against the force of deformation. In dynamic deformation methods, such as shot peening, the strain hardening is intensified by increasing the impact energy of tools (shot or balls). The impact energy depends on the mass (dimension) of the shot, or ball, and its velocity. With an appropriate choice of cold-working processing methods,

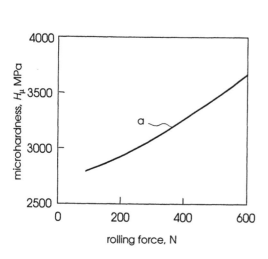

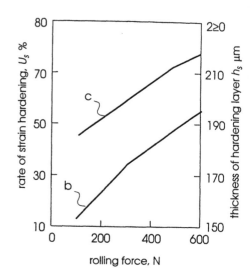

Figure 6.16 Hardness parameters versus deformation force in rolling of titanium alloys: (a) microhardness; (b) rate of hardness; (c) thickness of hardening layer

one can obtain a higher rate of strain hardening and a thicker hardened layer than obtained by mechanical processing. Even under small deformation forces, the microhardness in a surface layer may increase by 20 to 30 percent. This fact makes the cold-working methods preferable and allows us to dispense with heat treatment operations or to replace expensive metal alloys with cheaper materials. For example, a strain hardening to a depth of over 20 mm by rolling can replace quenching of long shafts, stocks, rods and spindles.

The process of cold working has restrictions. For example, beyond a certain hardening threshold, a destruction of surface layer may begin. The destruction manifests itself in reduced surface microhardness and appearance of microcracks. These can be guarded against through the properly chosen strain-hardening process.

6.2.4 Stability of strain hardening

Plastic deformation of the surface layer is characterized by a large number of crystal lattice defects, such as dislocations and vacancies, which demonstrate a deviation from equilibrium in the crystalline structure. The deformed structure of the surface layer is thermodynamically unstable, tending towards a more stable condition. The transition to a more stable state is accompanied by a reduction in the number of lattice defects, with a corresponding hardening. This transition can take place with external intervention, by force or heat. A working machine part may be subjected to static and cyclic loading, high temperatures and other factors. These effects may cause softening of the surface layer, affecting its other mechanical properties, such as fatigue strength.

Strain hardening of the surface layer and the duration of the hardening effect depend on the stability of the plastically deformed structure and the type and intensity of external factors. Experience has shown that when external loading and heating are absent, the hardness is stable and does not change with time. Thus, for example, tests with AISI 1045 steel parts (processed by turning, grinding and burnishing and allowed to age for 2 years) have demonstrated the absence of any detectable changes in surface microhardness and depth of strain-hardened layer. Also, application of external loads at room temperature has practically no effect on the strain-hardening parameters. According to data of Zobnin,[4] metal hardness and residual stresses in railway car axles (caused by rolling) remained stable during 500 million loading cycles. (This is equal to about 1,650,000 kilometers or 25 years in the operation of a railway car.)

Surface layer softening may be experienced by strain-hardened machine parts upon heating, depending on the level and duration of the heating. The softening process can be divided into two stages, recovery and recrystallization. The recovery which occurs spontaneously tends to restore original metal properties before mechanical processing. During the recovery stage the accumulated distortion energy in a crystal lattice stimulates atomic realignment, whereby the atoms assume lower-energy positions. The alignment eliminates local distortions (partially or completely) and reduces hardness and strength. In most metals the atomic realignment at room temperature is slow and, in order for the recovery to occur at a perceived speed, the temperature has to be raised by several hundreds degrees. The second stage is recrystallization which pertains to creation of new crystals with the same structure as the original but free from strain hardening caused by mechanical deformation. Recrystallization centers appear in clusters at grain boundaries and joints. New grains start from small nuclei, coming into contact with each other and covering the entire material. The new grains contain fewer dislocations than the deformed surface layer, therefore the strain hardening is partially or fully eliminated. Recrystallization usually occurs at temperatures $T \geq 0.4T_m$, where T_m is the melting point of the metal. The recrystallization temperature for steels and titanium alloys is in the range 400 to 550°C and for high-temperature resistant alloys 700 to 900°C.

Figure 6.17(a) presents the rate of strain-hardening (U_s) of alloy specimens, processed by different methods, plotted against the temperature of annealing. Figure 6.17(b) shows the variation of microhardness against the surface layer depth (Reference 3). The data show that up to a temperature of 700 to 750°C, heating has no practical effect on the rate of strain hardening and depth of strain-hardened layer. At higher temperatures, the surface microhardness, the rate of strain hardening and the strain-hardening depth sharply drop. Therefore, strain hardening of the surface layer, hardness and strength remain unchanged when heating a metal to a temperature within the range of recrystallization. Beyond that range, the surface layer hardness drops in proportion to the heating temperature.

6.2.5 Influence of surface layer on fatigue

The effects of surface layer on fatigue strength is illustrated in Figure 6.18, where *S-N* plots are shown for steel specimens produced by two kinds of turning operations: (a) forming a strain-hardened layer and (b) without strain hardening. (The former is achieved by changing the cutting velocity while other factors are kept constant, at a surface roughness $R_a = 1.5 - 2.0\ \mu\text{m}$.) As the figure shows, the metal in strain-hardened condition has a higher fatigue limit which means that an increase in surface layer hardness makes it possible to increase fatigue loading. Also, the strain-hardened surface layer blocks propagation of fatigue cracks and resists formation of the new ones. Such a layer may outweigh possible negative effects of surface roughness and other stress concentrations.

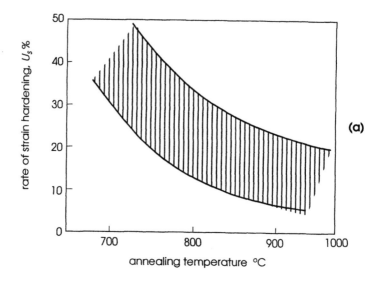

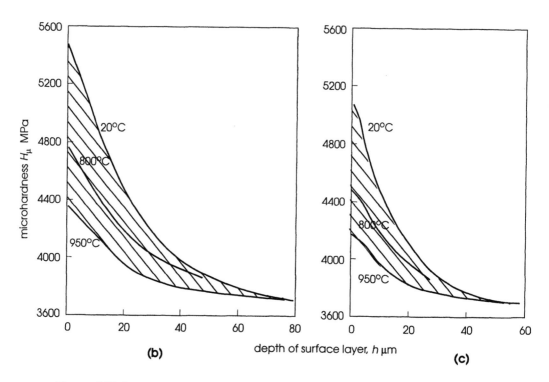

Figure 6.17 Strain hardening of the surface layer of a high-temperature resistant alloy: (a) rate of strain hardening versus annealing temperature; (b) microhardness versus depth after milling; (c) microhardness versus depth after grinding. (Data from Sulima, A. M., and M I Evstigneev, see Reference 3.)

Figure 6.19 shows fatigue fractures in two AISI 3318 alloy steel specimens: (a) a specimen without strain hardening and (b) a strain-hardened specimen. In the nonhardened specimen the crack begins on the surface, from where it propagates inside the specimen. In strain-hardened specimen the crack starts, typically, below the strain-hardened surface layer. The site of crack initiation has a round form from the center of which develops the crack. The distance from the surface to the crack initiation site increases with the thickness of strain-hardened layer. Fractographic analysis of fatigue fractures in parts with strain-hardened surface layers confirm that fatigue cracks do not originate on the surface, but under the strain-hardened layer. In strain-hardened parts the initiation of fatigue cracks occurs at higher stresses and after more loading cycles than in nonhardened ones.

The effect of strain hardening of the surface layer on fatigue strength depends on deformation of the layer and service conditions due to environment, temperature and loading. Figure 6.20 depicts the effects of strain-hardened layer thickness and rate of strain hardening on the fatigue limit of the AISI 1045 carbon steel. The fatigue limit increases in direct proportion to the thickness of the strain-hardened layer. Therefore, the mechanical processes that cause strain hardening of the surface layer also extend the fatigue strength. See Figures 6.12 and 6.13. Since the magnitude of strain hardening depends on the structural condition of the metal, lesser increase is found in steels with a sorbite structure, while in martensite and ferrite steels the magnitude is greater. The structural condition has the same effect on the fatigue limit. See Figure 6.21.

Surface decarburization

The processing of machine parts, depending on the treatment, can cause different defects in the surface layer which affect the fatigue strength such as surface decarburization, burns, overhardening and surface cracks. Surface decarburization may occur

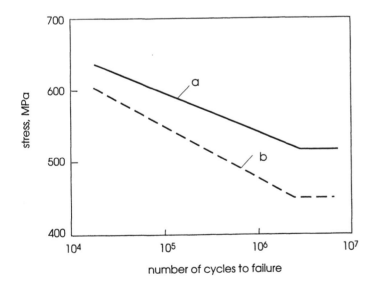

Figure 6.18 S-N curves for AISI 4335 steel, $R_c = 30–32$: (a) strain hardened by turning, (b) without strain hardening.

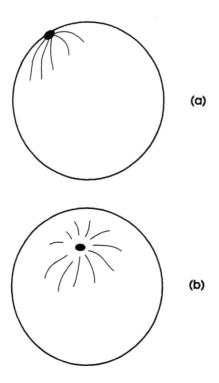

Figure 6.19 Crack generation in AISI 3318
steel, R_c=39-40: (a) without strain hardening;
(b) with strain hardening.

after the following: hot forming (forging, stamping), heat treatment (quench hardening) or in mechanical processing with high surface layer heating (grinding).
Decarburization reduces strain hardening of the surface layer and consequently the
fatigue strength. The thickness of decarburized layer measures several tenths of a mil-
limeter. Numerous data confirm the sharp decrease of fatigue strength due to
decarburization. Thus, the reduction in the fatigue limit in parts with "black" surface
after hot forming reaches a considerable drop: 20 to 40 percent for steels with ultimate
strength $S_u = 450–700$ MPa, 45 to 55 percent for steels with $S_u = 800–1000$ MPa, and
75 to 85 percent for steels with $S_u = 1800–2140$ MPa.

The extent of changes in the surface layer due to decarburization depends mainly
on the structural condition of the metal. Table 6.3 presents the results of decarburization
of a high alloy steel in different structural states after quench hardening.[5] The data confirm
the basic dependence of fatigue strength on the structure. In steels with a sorbite structure obtained in quench hardening, a low decarburization has practically no effect on the
fatigue strength. In steels with troostite structure the same decarburization also insignif-
icantly reduces the fatigue limit. The same level of decarburization in a martensite
structure, however, causes a sharp drop in the fatigue limit; higher decarburization causes
additional fatigue limit reduction (up to 50 percent). Thus, even insignificant decarburization of steels with a martensite structure causes the fatigue strength to drop sharply,
while in steels with a sorbite and troostite structures these changes are not evident.

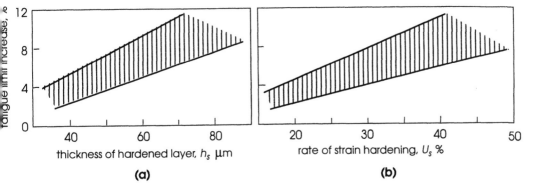

Figure 6.20 Increase of fatigue limit versus thickness of hardened surface layer and rate of strain hardening of AISI 1045 steel after turning.

Burns and microcracks

The local structural phase transformations (burns as well as microcracks) are possible when hardened steel parts are ground. (The nature of such defects and the conditions for their appearance were described above.) These grinding defects reduce considerably the fatigue life of machine parts and machine parts indicating such defects should be rejected. In most cases, however, after grinding the parts do not undergo such inspection and enter the operation with undetected defects.

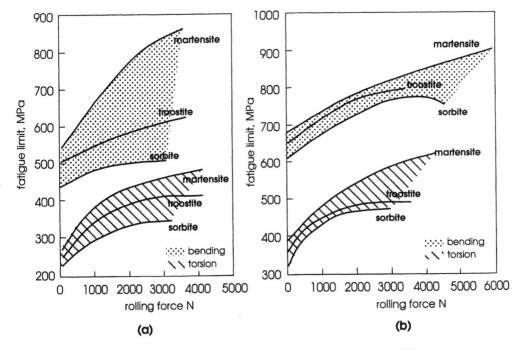

Figure 6.21 Fatigue limit versus rolling force for different structural conditions: (a) steel AISI 3318; (b) steel AISI 4345. (Data from Balter, M. A., see Reference 5.)

TABLE 6.3
Decarburization Rate and Fatigue Limit of Steel AISI 4345

Structure	HARDNESS HV Surface	Core	DECARBUNIZATION RATE CHEMICAL COMPOSITION Δ Carbon (percent)	Layer Depth (mm)	HARDNESS ΔHV	Layer Depth (mm)	ULTIMATE STRENGTH (MPa)	FATIGUE LIMIT (MPa)
Martensite	570	570	0	0	0	0	2080	670
	535	570	0.08	0.2	40	0.15	2080	430
	470	570	0.20	0.4	100	0.25	2080	330
Troostite	475	475	—	—	0	0	1530	650
	435	475	—	—	0	0.15	1530	620
Sorbite	420	420	—	—	0	0	1340	580
	380	420	—	—	40	0.15	1340	570

Source: Balter, M.A., Reference 5.

Let us follow through the conditions of defective specimens after grinding. An investigation was performed with parts made of steels AISI 5140 (R_c 54) and AISI W-1 (R_c 62) with different carbon contents. The parts were subjected to hardening and tempering under usual conditions. After the heat treatment the specimens were ground to three different cutting depths: a depth of 0.01 mm produced no detected burns; a depth of 0.06 mm without cooling caused a development of tempering burns; and a depth of 0.10 mm with cooling produced secondary hardening burns. The following analyses were conducted: microstructure analysis of the surface layer; X-ray analysis to determine residual austenite content; measurement of microhardness at the outer surface and at oblique cuts; checking for cracks by methods of color defectoscopy, magnetic control and microscopy.

The following conclusions were reached relating to the nature of surface defects. During the grinding process, tempering burns develop when the heat is quickly absorbed by the inner metal layers and the heating temperature of the surface layer does not exceed transformation temperature Ac_1 of the iron-carbon diagram. As this happens, the martensite structure of the hardened steel breaks down, creating a troostite-sorbite tempering structure. The hardness of the tempered layer decreases considerably—the hardness reduction is 24 to 28 percent for steel AISI W-1 and 15 to 20 percent for the steel AISI 5140. See Figure 6.22. The burns create in the surface layer a nonuniform structure with "spotted" hardness. When, in the process of grinding, the surface layer is heated beyond the temperature of phase transformation, secondary hardening burns develop. The secondary strain-hardened layer consists of hardened martensite and austenite, up to 40 to 60 percent, with a hardness up to 10–11 GPa and a thickness of 0.01–0.02 mm. A tempered layer with a reduced hardness, whose thickness is up to 0.1 mm, is located under the secondary strain-hardened layer. Rough grinding also causes surface cracks which can be detected by any of the available control methods.

In order to estimate the fatigue strength of parts with a structural damage caused by grinding, fatigue tests were conducted with specimens made of Cr-Si alloy steel (R_c 62). See Figure 6.23. The tests have shown a reduction of fatigue limit from 880 MPa to 560 MPa.

Surface damage due to strain hardening

When machine parts are processed by methods involving such factors as large deformation forces, repeated passes and small feeds (occurring in turning, burnishing and rolling) a damage of the surface is possible under certain conditions. The damage in the form of cracking or flaking strongly decreases the fatigue strength of machine parts. The main cause of this damage is an excessive strain hardening.

To understand the conditions leading to such damage, the effects of deformation on the surface layer and the occurrence of surface defects were studied. during a multipass rolling. The investigation was performed with ground steels of a martensite, troostite-martensite, sorbite and ferrite-pearlite structure, having a roughness $R_a = 0.6–0.8$ μm. The deformation forces varied within the limits of 50 to 500 N and the feeds were 0.02–0.05 mm per revolution. All the experiments were conducted under constant speed, 1 m/sec, to avoid surface heating. The multipass rolling was applied until the appearance of signs of cracks in the surface layer. The photographs of surface showed that an increase in the number of deformation cycles (repeated tool passes) changes the

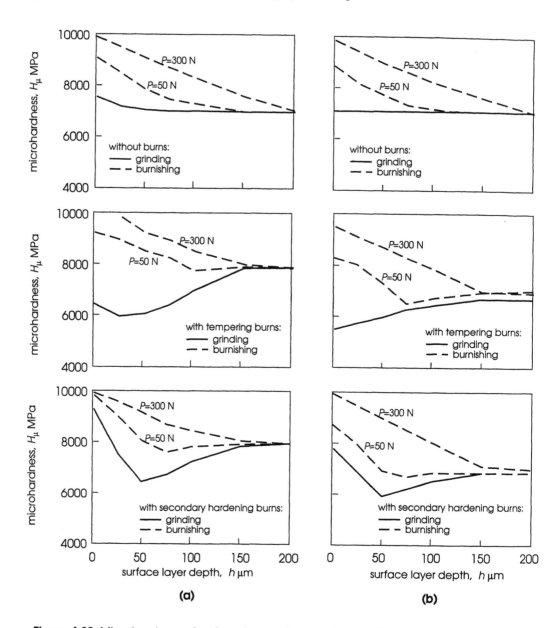

Figure 6.22 Microhardness of surface layer after grinding and burnishing: (a) tool steel AISI W-1, $R_c = 62-64$; (b) alloy steel AISI 5140, $R_c = 54$.

quality of the surface layer. The first tool pass left a more even surface, without any damage, as compared with the initially ground texture. Some repeated tool passes led to the appearances of isolated microcracks. With an increase in the number of passes the microcracks multiplied and grew, interconnecting and forming a continuous grid. The microcracks propagated into the material at an angle 20 to 45 degrees with respect to the surface. Metal flakes appeared and subsequently were detached from the surface. Valleys formed after the separation of flakes had a width of 0.05–0.25 mm and depth

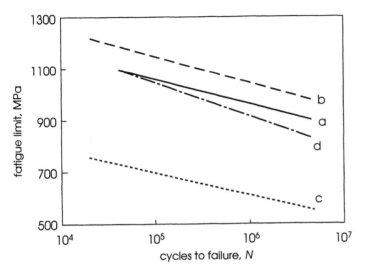

Figure 6.23 Fatigue limits of specimens of a Cr-Si alloy steel:
(a) grinding without burns, (b) grinding without burns and burnishing,
(c) grinding with burns, (d) grinding with burns and burnishing.

of 0.06–0.10 mm. The rolled surface, normally distinguished by a luster, became mat. (Figure 6.24 shows the variation in the surface finish and microhardness during burnishing.)

An analysis of surface condition, its roughness and microhardness as a function of deformation cycles, leads to a conclusion that the deformation of surface layer comprises three stages. The first stage is the stage of strain hardening (where the number of deformation cycles does not exceed 20) which consists of an intensive reduction of surface roughness height together with strain hardening of the surface layer. The second

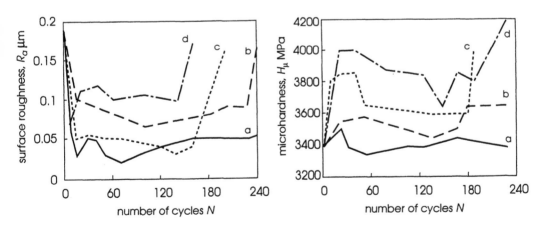

Figure 6.24 Surface roughness and microhardness of surface layer versus number of cycles for alloy steel AISI 5140, BHN 280: (a) load 50 N; (b) load 100 N; (c) load 200 N; (d) load 300 N.

stage, so called saturation stage (where the number of deformation cycles varies between 20 and 200), is characterized by stable surface roughness and microhardness of the surface. At this stage plastic shearing takes place, the density of dislocations and the magnitude of the crystal structure defects decrease and microcracks originate in the surface layer. In the third stage, the stage of surface damage, the integrity of the surface is destroyed, cracks and flakes appear and the surface roughness rises sharply.

The appearance of surface damages depends on several factors such as the contact pressure during processing, the number of deformation cycles and the structure and properties of the processed material. Figure 6.25 shows the relation between the number of deformation cycles that cause surface damaging and the contact pressure. The maximum damage that can occur during processing is related to the structure and hardness of the processed material: damage of strain-hardened steels of martensite structure occurs after a much higher number of deformation cycles than steels of pearlite or sorbite structure. When the hardness of the processed surface increases, its resistance increases and the damage appearance is in the direction of high loading cycles. See Figure 6.25.

The surface damage and cracking have a strong effect on mechanical properties. Therefore processing conditions that eliminate surface damaging must be chosen. The condition of an undamaged surface can be presented as

$$B_{\lim} > B = p_o N^m \qquad\qquad [6.11]$$

Table 6.4 presents values of $B_{\lim}$ and m for a number of materials. For the materials listed in the table, $B_{\lim} = 1.1–1.3$. Equation (6.11) together with Table 6.4, allow us to choose processing conditions to reduce surface damage and prevent fatigue strength decrease in the listed materials and materials with similar mechanical properties.

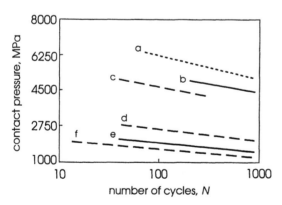

Figure 6.25 Number of deformation cycles damaging the surface versus contact pressure: (a) bearning steel, $R_c = 64$; (b) carbon steel AISI 1045, $R_c = 45$; (c) alloy steel AISI 5240, $R_c = 56$; (d) alloy steel AISI 5140, BH = 300; (e) carbon steel AISI 1045, BHN = 200; (f) alloy steel AISI 5140, BHN = 200.

TABLE 6.4
Values of B_{lim} and m Parameters

Material	Hardness	B_{lim}	m
Tool steel W-1	$R_c = 64$	1.25	0.10
Bearing steel UNS G52986	$R_c = 64$	1.10	0.08
Carbon steel AISI 1045	$R_c = 56$	1.13	0.07
Chromium steel AISI 5140	$R_c = 55$	1.10	0.08
Alloy steel AISI 4335	$R_c = 52$	1.13	0.08
Chromium steel AISI 5140	$R_c = 50$	1.20	0.08
Chromium steel AISI 5140	BHN = 280	1.15	0.06
Chromium steel AISI 5140	BHN = 200	1.35	0.09
Carbon steel AISI 1045	BHN = 200	1.25	0.06

Effect of high temperatures

As shown, strain hardening in the surface layer is reduced or totally eliminated in parts operating at high temperatures that equal, or exceed, the recrystallization temperature of the metal. These conditions occur in a number of systems, such as jet propulsion engines and power plants. Reduced hardness under high-temperature conditions reduces the fatigue limit of the metal.

Table 6.5 presents results of fatigue tests at different temperatures of specimens made of stainless steel, strain-hardened and nonhardened.[6] The data in the table show that, as the testing temperature increases, the fatigue decreases for all specimens. There is, however, an advantage for strain-hardened metals which decreases with temperature (at 600°C the advantage becomes small). A fatigue fracture analysis of specimens shows that in polished (nonhardened) specimens fatigue cracks started from the surface at all temperatures. In strain-hardened specimens at temperatures of 250°C and 450°C the cracks started under the surface, demonstrating the effect of the strain-hardened layer. At 600°C the cracks in strain-hardened specimens started on the surface, indicating the strain hardening has practically vanished. Thus at high service temperatures that exceed the recrystallization temperature of the metal, the strain hardening of the surface layer is practically eliminated and the fatigue strength does not increase.

At very high temperatures above the temperature of recrystallization, strain hardening of the surface layer reduces the fatigue strength. Fatigue tests with specimens made of high-temperature resistant alloys performed at 800°C and 900°C indicate that higher strain hardening causes a decrease of fatigue strength. See Figure 6.26. To reduce the adverse effect of strain hardening in parts subject to high temperatures and cyclic alternating loading, one should use the processes that will remove the most deformed upper layers of the metal (electropolishing, electrochemical machining, and polishing) or that will eliminate the surface layer strain hardening by heat treatment.

Strain hardening of a surface layer also adversely affects parts operating under alternating thermal conditions (thermal fatigue). Table 6.6 lists test results of high temperature resistant nickel alloy specimens subjected to cyclic heating and cooling.[7] The results show that fatigue strength depends on the maximum heating temperature and

TABLE 6.5
Fatigue Test Results for Stainless Steel 321

Test Temperature (°C)	Processing	FATIGUE LIMIT	
		(MPa)	Relative to polished
	Polished	201	1.0
250	Strain hardened (P = 170N)	303	1.51
	Strain hardened (P = 340N)	365	1.83
	Polished	192	1.0
450	Strain hardened (P = 170N)	256	1.33
	Strain hardened (P = 340N)	320	1.67
	Polished	156	1.0
600	Strain hardened (P = 170N)	168	1.08
	Strain hardened (P = 340N)	191	1.23

Source: Khvorosukhin, L.A., see Reference 6.

temperature amplitude. At a maximum cycle temperature of 750°C, under the recrystallization temperature of the high-temperature resistant alloy, there is practically no difference between the number of cycles to failure of polished and strain-hardened parts. At a maximum cycle temperature of 900°C, exceeding the recrystallization temperature, hardened parts fail 1.8 to 2.1 times faster.

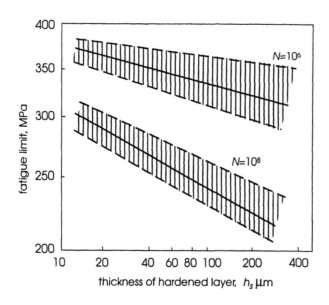

Figure 6.26 Fatigue limit versus thickness of hardened surface layer for high-temperature resistant nickel alloy; test temperature 800°C. (Data from Sulima, A. M., and M I Evstigneev, see Reference 3.)

TABLE 6.6
Thermal Fatigue Test Results of
High Temperature Resistant Nickel Alloys

Heat Cycle (°C)	NUMBER OF CYCLES TO FAILURE		
	Polished N_p	Rolled N_r	N_p/N_r
200–750	15400	15000	1
450–900	4033	2240	0.55
200–900	740	350	0.47

Source: Kishkin, S. G., and A. M. Sulima, see Reference 7.

Concerning the effect of the surface layer and its strain hardening on fatigue strength of machine parts, the following main conclusions may be drawn:

(a) Fatigue strength is strongly affected by the structural state of the material and especially by the structural state of the surface layer. Martensite structures are most fatigue resistant. Troostite and sorbite are the least fatigue resistant.

(b) In parts operating below the temperature of recrystallization of the metal surface, strain hardening of the layer due to mechanical processing increases significantly the fatigue strength of the part by increasing the metal hardness of the surface layer and moving the site of fracture from the surface to internal layers.

(c) In parts operating at very high temperatures that are above the temperature of recrystallization, the surface layer hardness obtained by mechanical processing reduces the fatigue strength of the part.

(d) Surface defects formed in manufacturing (surface layer decarburization, burns, cracks, and surface layer overhardening) significantly reduce the fatigue strength of parts.

6.3
RESIDUAL STRESSES

Residual stresses are defined as those stresses that exist in a machine part without external loading (forces or temperature) being present. These stresses usually occur in manufactured parts as a result of irregular elastic-plastic deformations in welding, heat treatment and mechanical processing. Due to their nature the residual stresses are in equilibrium (self-balance).

6.3.1 Classification

There are, in fact, two distinct kinds of residual stresses with an inherent difference between the two: macrostresses and microstresses. Macrostresses are those residual stresses which are in equilibrium within macrodomains, covering volumes comparable in size to the part; they concur with the concept of isotropic material. Microstresses are crystal stresses within single metal grains or groups of grains in equilibrium within volumes comparable with dimensions of the grains. These include submicroscopic stresses which relate to distortions in atomic lattices of crystals. (To date microstresses have not been adequately researched.) Macrostresses have been widely investigated, including the reasons for their occurrences and how they affect the strength and other properties of metals. The macrostresses are listed in technical specifications of parts and are inspected in surface layers after manufacturing.

Since the discussion that follows is limited to residual macrostresses only, all mentions of residual stresses from now on refer to macrostresses. In general, the residual stresses are either tension stresses or compression stresses. Residual tension stresses are detrimental, since they reduce fatigue strength and impair wear and corrosion resistance of machine parts. Compression stresses, to the contrary, are believed to be beneficial. In the manufacture of machine parts the residual stresses are grouped according to the shape and trajectory of the processing tool: tangential, in the direction of the displacement of the tool; axial, perpendicular to the tool displacement; and radial, perpendicular to the processed surface.

6.3.2 Origin of residual stresses

The creation of residual stresses is caused by elastic-plastic deformation, which takes place in the contact area between the tool and workpiece and is aided sometimes by high temperatures. There are number of factors attributed to the origin of residual stresses, tension and compression, as we follow the processing:

Factor 1. Under the action of the tool on the processed metal surface three domains are formed in the surface layer: plastically deformed metal, elastically deformed metal, and unaltered initial state. The plastically deformed metal has lower density and therefore higher specific volume (0.3 to 0.8 percent more compared to the initial condition). The internal layers prevent the increase of volume of plastically deformed metal, resulting in residual compression stresses in the outer layer and tension stresses in inner layers. (Note: the higher specific volume above is taken from experimental data that seems to contradict the mathematical theory of plasticity, with assumption of specific volume being constant, see Chapter 2. Therefore it should be considered as an experimental data disregarded by the plasticity theory.)

Factor 2. As the tool moves along the processed surface, it stretches crystalline grains of surface layer, subject to elastic and plastic tensional deformation. This process is aided by friction of the tool on the processed surface. After the tool leaves the surface, the plastically stretched layer, bonded to inner layers, acquires residual compression stresses. Accordingly, the inner layers develop residual tension stresses for counterbalancing.

Factor 3. The heat released in the processing area instantly heats a thin surface layer to a high temperature, increasing the specific volume. On termination of the tool action a rapid cooling of the surface layer takes place causing a compression. The cooling is hindered by unheated inner layers. As a result, residual tension stresses are present in the surface layer, while compression stresses exist in the inner layers. This is most pronounced in high-velocity machining of materials with low thermal conductivity as, for example, in grinding of parts made of titanium alloys.

Factor 4. The high pressure and temperature during deformation cause structural-phase transformations in the surface layer. This is accompanied by a change in volume. When the specific volume of the surface layer increases, residual compression stresses develop. When the specific volume decreases, residual tension stresses appear. In processing with strong forces, for instance, a transition from austenite into martensite with a lower density takes place and residual compression stresses develop. On the other hand, in high-speed processing at high temperatures, the martensite is transformed into troostite or sorbite with a lower specific volume and residual tension stresses develop. In surface hardening of parts made of titanium alloys with $(\alpha + \beta)$ structure, a partial transformation of body centered cubic lattice β-Ti into hexagonal lattice α-Ti embrittles the metal, increasing its volume and producing residual compression stresses.

Any one of the above factors, by being dominant, dictates the nature and magnitude of residual stresses in the surface layer. A change in the processing methods will alter the deformation, force and temperature patterns in the processed area, creating different residual stresses. Processing methods where the force action predominates, create increased residual compression stresses and reduced residual tension stresses. Contrary results are achieved by processing methods with a predominant heating factor, generating residual tension stresses and reducing residual compression stresses.

6.3.3 Measurement of residual stresses

The residual stresses may be determined by either analytical or experimental methods. Analytical methods, developed by a number of authors, allow us to compute the distribution of residual stresses in the surface layer on the basis of the following factors: mechanical properties of the processed material, shape and dimensions of the part, and the loading condition. Because of the complexity of these methods, their application is limited. Often, important data is unavailable, such as strength and heat transfer properties of the material in a transformed surface layer or the intensity of heat sources. Or, the predicted residual stresses are inaccurate. In view of the above, the residual stresses are determined mostly by experimental methods.

The experimental methods are divided into destructive and nondestructive methods. In the destructive methods, the residual stresses are derived from measurements of deformations occurring after cutting the inspected part. The methods are based on the assumption that cutting the part along a given section is equivalent to applying stresses to the cut surface which are opposite to the residual stresses. These opposite stresses either cause deformations in the part or reaction forces in holding devices. By measuring the deformations or reaction forces the magnitude of residual stresses is derived. The destructive methods may be subdivided further into those that completely destruct the inspected part and those causing a partial destruction only. (Almost all destructive methods in use today involve complete destruction.)

Nondestructive experimental methods are based on measurements of electromagnetic, optic and other physical phenomena in the residual stress field. Residual stresses are measured in the tangential, axial and radial directions. Experience shows that since radial stresses are usually negligible, only the tangential and axial components are considered.

Sectioning inspection method

The method pertains to sectioning the tested part and inspecting a removed specimen in a form of a plate, ring, or prism. The specimens are cut by methods requiring minimum force and lowest heating to cause minimum changes to the residual stresses.

These include electric discharge machining, electrochemical machining, as well as others. The specimens are placed in a measuring device and are subjected to a continuous surface layer removal in the location of interest. See Figure 6.27. The stressed surface layers are cut away, one layer at a time, changing the field of stresses and causing the specimen to deflect. The deflection is continuously measured and recorded. The removal of metal layers from the investigated surface is usually done by electrochemical polishing or etching. At the same time other surfaces are insulated by a chemically resistant lacquer.

Plate specimen   Consider the inspection of a plate specimen. If the specimen, after this procedure, becomes convex toward the removed stressed surface layer, this indicates that there are residual compression stresses. If the specimen becomes convex in opposite direction, then tension stresses prevail. See Figure 6.27. (Similar phenomena happen in the ring specimen: it will contract when residual compression stresses are present at the outer surface, and expand when tension stresses are present)

The magnitude of the residual stresses in a removed layer may be computed from the specimen's deformation. The stress in a layer comprises three components

$$\sigma = \sigma_1 + \sigma_2 + \sigma_3 \qquad\qquad\qquad [6.12]$$

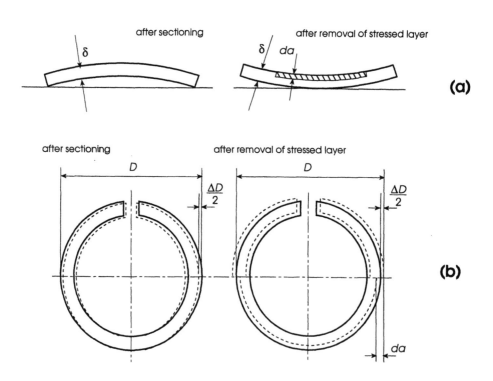

Figure 6.27 Specimen for measurement of residual stresses: (a) prismatic specimen; (b) ring specimen.

where σ_1 is the stress removed from the layer when cutting the specimen, σ_2 is the stress removed when slicing the specimen, and σ_3 is a correction stress which takes into consideration changes caused by the removal of previous layers. The σ_3 term is very small in comparison to the total sum. In turning operations, for example, σ_3 amounts to only about 2 to 3 percent of the total stress and therefore may be neglected. In plate specimens, according to computations, at depths that are one order of magnitude less that the specimen's thickness, the σ_1 component does not exceed 5 to 6 percent of the total stress and therefore it can also be neglected. As a rule, the σ_2 component predominates in all cases.

Ring specimen   In a ring specimen the tangential residual stresses σ_1 and σ_2 can be derived from the following expressions

$$\sigma_1 = \frac{E\alpha}{2\rho} \cdot \frac{Z' + \dfrac{\delta^2}{12\rho}}{Z' + \rho} \Delta D \qquad [6.13]$$

and

$$\sigma_2 = \frac{Ea^2\alpha}{3\left(D_b + a\right)^2} \frac{dD}{da} \qquad [6.14]$$

with following designations:

E = modulus of elasticity,
ρ = mean circumference radius of the ring,
Z' = distance from mean circumference to the middle of removed layer,
δ = wall thickness of the ring,
ΔD = change in the inner diameter after cutting in mm;
a = distance from the inner surface to the middle of removed layer,
D_b = inner diameter of the ring,
dD = change of inner ring diameter after removal of the layer,
da = width of removed layer.

Coefficient α denotes a function of Poisson ratio

$$\alpha = \frac{1}{1 - v^2} \qquad [6.15]$$

which takes into consideration the plane condition.

Prismatic specimen   For determination of axial residual stresses, prismatic specimens are used and σ_1 and σ_2 are computed from the following expressions

$$\sigma_1 = \frac{2E\left(a - \dfrac{\delta}{2}\right)}{b^2} f \qquad [6.16]$$

and

$$\sigma_2 = \frac{Ea^2}{3b^2}\frac{df}{da} \qquad\qquad [6.17]$$

where the designations are

  b = half-distance between supports in measuring specimen's deflection,
  f = deflection of the specimen,
  df = change in deflection after layer removal.

Measuring devices based on the sectioning method usually record the deformation of the specimen, perform differentiation of the deformation curve, and provide diagrams of the residual stress distribution. The method is reliable, providing accurate and quick results. Its main shortcoming is the destruction of the investigated machine part. There are a number of devices based on this method which differ in handling techniques.

X-ray diffraction method

The X-ray diffraction method for determination of residual stresses is based on measuring crystal lattice distortions in the material caused by residual stresses. Interplanar spacing in a crystal lattice is proportional to the magnitude of residual stresses. Measurement of interplanar spacing indicates the magnitude of residual stresses. The measured item is the angle of diffraction of an X-ray beam. The measurement is performed using an X-ray diffractometer. The advantage in this method is that it is nondestructive and thus the evaluated part remains undamaged. Another advantage is that the inspected part is not subjected during the measurement to conditions that might distort the field of residual stresses.

The X-ray diffraction method only measures surface stresses, since the X-ray beam penetrates the material to a depth of less than 0.025 mm. If a stress gradient in the surface layer is to be determined, thin layers must be removed from the measured surface, whereby the method then becomes a destructive one. The stress gradient is obtained from X-ray diffraction measurements of successively removed layers. The layers are removed by methods that do not affect the state of stress, for example, electrochemical polishing. The disadvantages to this method are its complexity, expensive equipment and low accuracy. The X-ray diffraction method is best used in the determination of residual stresses in small intricately shaped parts.

Other inspection methods

Several other methods for measuring residual stresses, with a lesser acceptance, are also presented below.

Electromagnetic method   The method is based on measuring the magnetic permeability of ferromagnetic materials. Magnetic permeability is an almost linear function of mechanical stresses and as such its measured values can be used to determine the residual stresses. In order to derive the complete stress gradient the metal should be stripped layer by layer. To perform the measurements of residual stresses in parts made of nonferromagnetic materials, it is necessary to apply a ferrolacquer to the inspected surface. The advantages of this method are its simplicity, high productivity, and a non-destructive estimation of residual stresses in the thin surface layer. Its disadvantage is that the electromagnetic properties depend not only on the residual stresses but also on the chemical and structural homogeneity of the material, heat treatment and other factors, all of which reduce the reliability of results.

Ultrasonic method   The method is based on measurement of the speed of sound in the metal. It assumes that the speed of sound in the metal varies linearly with stresses.

6.3.4 Effect of processing methods on residual stresses

The dominant causes creating residual stresses dictated by the processing have been described in Section 6.3.2. Depending on processing methods, these causes may have a greater or lesser effect on the character of residual stresses. The effects can come as a result of a combined action of several causes. By altering processing conditions, both the character and magnitude of residual stresses in the surface layer of a part can be changed. Let us consider the major processing methods as the source of residual stresses in machine parts.

Chip removal processing

In chip removal processes the predominant factor in formation of residual stresses is the plastic deformation of the surface layer. The effect of heating is also important. In processing metals with low ductility, discontinuous chips are produced that enhance the development of residual compression stresses. In processing ductile metals, on the other hand, continuous chips are produced with the appearance of residual tension stresses.

The turning operation usually produces residual tension stresses in the surface layer of 300–600 MPa in a depth of 0.05–0.20 mm, depending on the conditions. Milling, broaching and planing operations produce both tension and compression stresses. In most cases the maximum tension stresses are located at the surface, at a depth up to 0.03–0.04 mm. Consider the main operating parameters, such as the cutting velocity, feed, and geometry of the tool, and these affect the residual stresses.

The cutting velocity has a pronounced effect on the magnitude of residual stresses and their distribution. As the cutting velocity rises from 2 to 75 m/min, both tangential and axial residual tension stresses develop in the surface layer. See Figure 6.28(a). The magnitude of tangential residual stresses grows up to 750 MPa with increased velocity. Axial residual stresses grow similarly under the same conditions. In cutting ductile metals with residual tension stresses, increased cutting velocities will produce thermal tension stresses with consequent increased residual tension stresses. Processing low-ductile metals, increased cutting velocity produces thermal tension stresses that reduce the existing residual compression stresses and may even produce residual tension stresses.

The other important parameter, the feed of the cutting tool, increases proportionally the plastic deformation in the surface layer. Processing ductile metals, increased feed increases residual tension stresses. See Figure 6.28(b). When low-ductility metals are processed, increased feed increases residual compression stresses which penetrate deep into the surface layer.

Consider the geometry of the cutting tool. The strongest effect on surface layer deformation and upon the residual stress field is caused by the face angle. When the

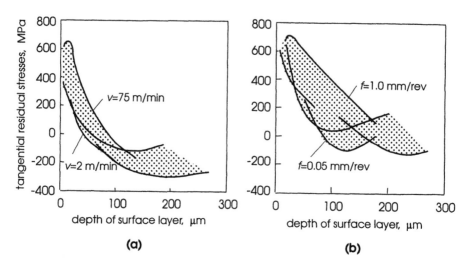

Figure 6.28 Distribution of tangential residual stresses in the surface layer of high-temperature resistant nickel alloy after cutting: (a) versus velocity of cutting tool; (b) versus feed.

front face angle is reduced the tool becomes dull, enhancing plastic deformation. This, in turn, increases residual compression stresses and reduces tension stresses. See Table 6.7. Processing ductile metals reduces residual tension stresses, while in low-ductile metals residual compression stresses increase. Tool wear has a profound impact on residual stresses, increasing the contact area between tool and workpiece, enlarging the nose radius of the cutting tool, and reducing the face angles to negative values—all of which enforces the wedge action of the tool. In processing stainless steel 321 residual tension stresses decrease with increased tool wear at the surface. See Table 6.7. In all cases, the depth of residual stresses significantly increases with increasing tool wear.

With the cutting force being the deciding factor in forming residual stresses, the effect of cooling can be treated by considering the changes between strain hardening and softening of the surface layer. With increased deformation, cold work and yield strength increase, producing larger residual stresses. High cutting temperatures and extended heating duration cause considerable softening of the metal which reduces

TABLE 6.7
Influence of Cutting Tool on Residual Stresses in Milling, Stainless Steel 321

Tool Angle φ Tool Wear h	Surface Residual Stress (MPa)	Layer Depth (mm)
$\varphi = 10°$	−20	0.15
$\varphi = 20°$	200	0.10
$h < 0.1$mm	500	0.15
$h = 0.25$mm	300	0.20
$h = 0.5$	100	0.30

residual stresses. In processing metals that are not prone to structural and phase transformations, the cooling reduces the possibility of softening, thereby enhancing the residual stresses formed by plastic deformation. For example, during processing of a titanium alloy by milling with no cooling, the maximum residual compression stresses reach 300–380 MPa and, if cooled with a 5 percent emulsion, they increase to 500–580 MPa. When milling stainless steel AISI 321 at speeds of $v = 25-86$ m/min, cooling with a 5 percent emulsion increases residual tension stresses by a factor of 1.5 to 2. At higher cutting velocities the factor is even greater. Similar data were obtained for high-temperature resistant alloys. When cutting tools are used with negative face angles and at higher velocities (that means, when the cutting temperature becomes higher) the increase of residual stresses in processing with cooling is more substantial. In high-temperature processes (when thermoplastic deformations in the surface layer become substantial and structural and phase transformations take place) the effect of cooling is complicated and may cause both a reduction as well as an increase of residual stresses.

Thus, residual tension and compression stresses develop in the surface layer in chip removal processes, depending on the type of material and processing conditions. By changing the tool's geometry, velocity and feed, and applying cooling, the residual stresses may be controlled over wide limits.

Abrasive processing

Let us start with the main abrasive process, grinding, characterized by strong thermal effects. The grinding process causes high stress and deformation rates, leading to high temperatures in the range of 600 to 800°C (reaching sometimes up to 1000 to 1600°C) while heating and cooling rates of the surface layer may go up to hundreds or thousands of degrees per second. Under these conditions, the known causes leading to residual stress development become significant, affecting their distribution and magnitude. Most often generated residual stresses are tension stresses. See Figure 6.29. The action of the abrasive wheel upon the metal, which becomes plastic at high temperatures, produces continuous chips that cause tangential residual tension stresses. At the high temperature, there are additional residual tension stresses from structural and phase transformations reducing the metal volume (secondary strain-hardening).

The residual tension stresses at high temperatures are caused by the thermoplastic deformation. The onset temperature at which it begins (Θ_i) can be computed from the expression

$$\Theta_i = \frac{\sigma_{yp}}{\alpha E} \tag{6.18}$$

where σ_{yp} is the yield point of the material in the processed area, α is the thermal coefficient of expansion, and E is Young's modulus. When the temperature in the processed area exceeds the onset temperature, residual tension stresses develop. The values of onset temperature for different metals are listed in Table 6.8.

The residual stresses depend on temperature and duration of heating during processing. The following conditions increase the temperature and, consequently, the residual tension stresses in the surface layer: reduced heat conduction of the processed material, inadequate cooling, increased rotational velocity of the grinding wheel, dull

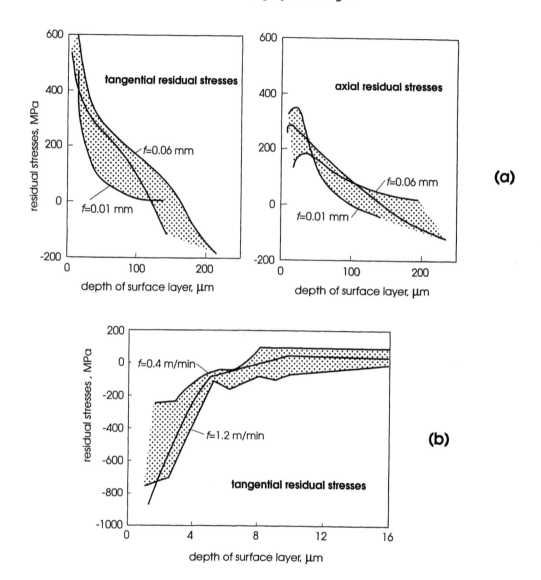

Figure 6.29 Residual stresses in the surface layer of high-temperature resistant nickel alloy: (a) after grinding; (b) after superfinish. (Data from Sulima, A. M., and M I Evstigneev, see Reference 3.)

wheel, increased cutting depth, and reduced velocity of feed (circular or longitudinal). The opposite conditions will reduce residual tension stresses and increase residual compression stresses. The most effective means for reducing temperature is to increase the velocity of the part (reduced duration of heating) and to cool effectively the immediate contact area.

Figure 6.29(a) shows the typical plots of tangential and axial residual stresses in the surface layer caused by grinding. Maximum tension stresses occur at the surface, rapidly diminishing with depth and turning into compression stresses at a depth of 0.10–0.20 mm.

TABLE 6.8
Temperatures of Thermoplastic Deformation

Material	Θ_i (°C)
Carbon steels	100–180
Stainless steels	80–100
High temperature resistant Cr-Ni steels	400–500
High temperature resistant nickel alloys	230–340

Axial residual stresses caused by grinding are also tension stresses. They are smaller in magnitude than tangential stresses and their penetration depth is shallower. In general, the magnitude and the penetration depth of residual stresses vary within wide limits corresponding to the characteristics of the abrasive wheel and grinding conditions; the residual stresses are mostly tension stresses.

Other abrasive processes, such as polishing, lapping, honing and superfinish, are performed at low velocities and pressures. Since the thermal effects are small, the plastic deformation becomes the dominant factor, causing residual compression stresses in the surface layer. The residual compression stresses reach 750–850 MPa, in superfinish operations of parts made of hardened steel AISI 1045, for instance, penetrating to a depth $h = 0.005-0.015$ mm (Reference 3). See Figure 6.29(b). They have the same values in honing with penetration depth reaching 0.030 mm.

Cold-work processing

During cold-work processing, such as shot peening, rolling and burnishing, the surface layer of the processed part is subjected to an intensive plastic deformation, while the heating is low. When hardened steels with martensite structure are processed, the phase transformations that take place are reduced to almost complete residual austenite dissociation with its conversion to martensite. These effects cause the development of residual compression stresses in the surface layer.

Figure 6.30 shows typical diagrams of residual stresses resulting from cold-work processing. As shown in Figure 6.30, significant compressive stresses are formed in the surface layer that are close to yield point of the material. In rolling and burnishing the compressive stresses reach to a depth of 0.15–0.30 mm, depending on the material and processing conditions. Maximum tangential stresses occur not at the surface, but at a certain depth. Maximum axial stresses are at the surface, gradually decreasing toward the internal layers. Below the depth of 0.1 mm the tangential and axial residual stresses become essentially equal.

In shot peening and tumbling the distribution of residual compression stresses is roughly the same in all directions. In processing the surface with shot over 1 mm in size, however, maximum stresses were observed in depth proportional to diameter of the particle. See Figure 6.30(c). When using microballs 0.16–0.20 mm, the maximum compression stresses were observed at the surface. The latter phenomenon improves the fatigue strength of a machine part. Also, in a hardened surface layer of 0.05–0.1 mm thickness processing with rotating round metallic brushes produces maximum compression stresses at the surface.

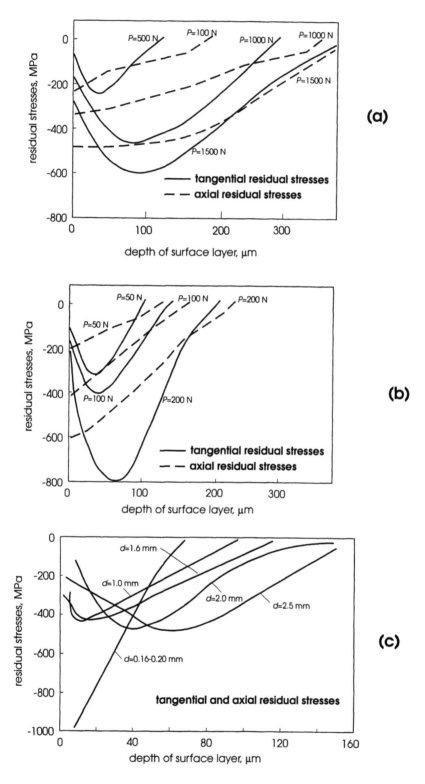

Figure 6.30 Residual stresses in the surface layer after cold-working: (a) rolling; (b) burnishing; (c) shot peeninng. Rolling and burnishing data pertain to tool steel T1, shot peening data pertain to titanium alloy.

In cold-working methods, as noted above, the acting forces have a strong effect on formation of residual compression stresses. The contact pressure affects the intensity of deformation and subsequently controls the level of residual stresses. By increasing the deformation and impact forces and prolonging the processing time, higher stresses are obtained. The increase of residual stresses is restricted by a deformation limit, called "the saturation threshold" of the material.

Tests for ten different steels were conducted, in order to obtain a correlation between the residual stresses and the contact pressure. The results are plotted as shown in Figure 6.31. To eliminate the hardness effect, maximum residual stresses, $\sigma_{r\max}$, were presented in a nondimensional form ($\sigma_{r\max}/HV$). The derived correlation can be approximately expressed by the linear equation

$$\frac{\sigma_{r\max}}{HV} = 0.4\left(p_o - 0.4\right)$$

[6.19]

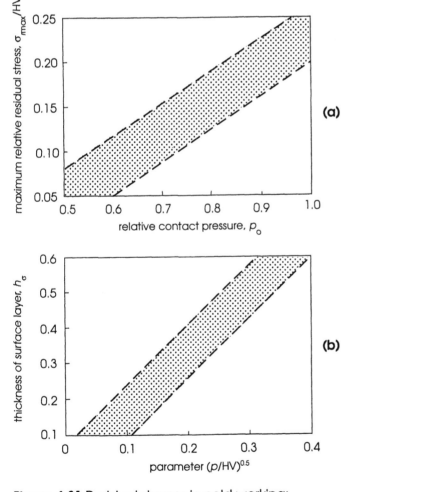

Figure 6.31 Residual stresses in cold working:
(a) maximum residual stress versus contact pressure;
(b) thickness of surface layer containing residual
stresses versus parameter $(p/HV)^{0.5}$.

where

$$p_o = \frac{p}{HV} \qquad [6.20]$$

HV is Vickers hardness, p is the contact pressure between the tool and the workpiece, and p_o denotes relative contact pressure. The thickness of the surface layer, containing residual compression stresses (derived from Figure 6.31), is expressed by the equation

$$h_\sigma = 1.75\left(\frac{p}{HV}\right)^{0.5} \qquad [6.21]$$

where P is the processing force.

Cold-work methods are used in final processing, after the cutting (chip removal and abrasive) processes. Investigations have shown that the magnitude and depth of penetration of residual stresses caused by cold work are independent from the preceding operations. Cold-work processes produce residual compression stresses in the surface layer, beneficial to a machine part, as explained below.

6.3.5 Relaxation of residual stresses

Residual stresses generated by processing may change during the operation of the part. The alteration depends on many factors, such as loading, temperature, work duration and the surface layer condition. The surface layer with a large number of dislocations due to processing is subject to a thermodynamic instability, in particular, at high temperatures. The instability is due mainly to the changes in crystalline structure and to the disturbed energy equilibrium. During the operation of the part a restoration of equilibrium takes place, resulting in a more stable condition with lower free energy. Together with the change of structural parameters a relaxation of residual stresses takes place, prompted by the diffusion processes.

The relaxation of residual stresses can be measured by the expression

$$U_\sigma = \frac{\sigma_{ri} - \sigma_{rt}}{\sigma_{ri}} \cdot 100\% \qquad [6.22]$$

where σ_{ri} is the initial magnitude of residual stresses and σ_{rt} is the magnitude of residual stresses in the instant t. The relaxation and the rate of change grow proportionally to the temperature, duration, and the deformation level of surface layer. The initial relaxation rate is high, decreasing progressively.

In structural steels subject to bending at room temperature, with stresses near the yield point, over long periods of time (up to 50,000 hours) no relaxation was noticed. According to tests, steel AISI 9230 kept outdoors for 2,000 hours after burnishing did not display any signs of relaxation. At high temperatures a relaxation of residual stresses takes place. Figure 6.32 presents plots of residual stresses in the surface layer of a high-temperature resistant nickel alloy, ground and heated to different temperatures (see Reference 3).

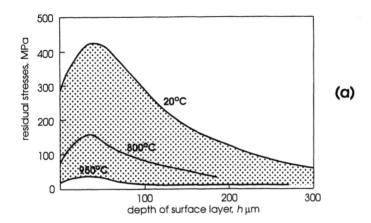

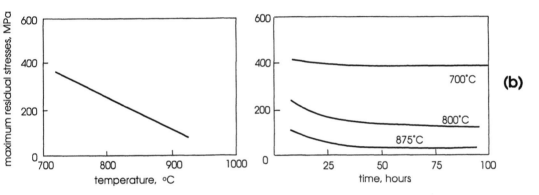

Figure 6.32 Relaxation of residual stresses in a high-temperature resistant alloy: (a) residual stress distribution in the surface layer; (b) maximum residual stresses versus temperature and time. (Data from Sulima, A. M., and M I Evstigneev, see Reference 3.)

At temperatures about 700 to 750°C no effect on residual stresses was observed, while at temperatures 950 to 975°C, after 2 hours, they were almost nonexistent. A sharp drop in residual stresses at given temperatures is also noticed in Table 6.9. Plots in Figure 6.32(b) reveal a sharp drop in the magnitude of residual stresses in the first 24 hours, after which no stress relaxation was noticed; although absolute relaxation is never

TABLE 6.9
Temperature of the Decrease of Residual Stresses

Alloy	Temperature (°C)
Titanium alloys	400–450
High-temperature resistant steels	450–500
High-temperature resistant nickel alloys	650–800
Superalloys	800–900

achieved, there always remains a small scatter of residual stresses. The factor of temperature has a stronger effect on relaxation than the factor of time. When testing cold-worked turbine blades of a gas-turbine engine, made of a high-temperature resistant alloy, no residual stress relaxation was noticed after 300 hours of operation at 370°C. This temperature is 100 to 120°C lower than the temperature at which residual stresses experience a sharp drop.

Application of external loading has a pronounced effect on residual stresses and their relaxation which depends on the condition of the surface layer: either elastic or plastic. If the effective stress, resulting from combination of residual and applied stresses, exceeds the yield point, the surface layer becomes plastic and relaxation takes place. Therefore, in materials with low yield point, or at higher stress magnitudes caused by high external loading, the residual stresses are less stable. A plastic deformation of one percent in the surface layer causes a near complete stress relaxation. Considering cyclic loading, most residual stresses diminish during the initial period. At smaller loads, with effective stress below the yield point, the relaxation will not occur.

In summary, residual stresses in a surface layer of a part are either retained indefinitely or relax, depending on the working conditions (such as heating and static and cyclic loading). The rate of relaxation depends on heating temperature and the stress level.

6.3.6 Effect of residual stresses on fatigue

In a working machine part the residual stresses affect the fatigue life, depending on whether they are tensional or compressive. Figure 6.33 shows different effects of residual stresses in cyclic loading with corresponding applied cyclic stresses. The applied cyclic stresses vary between tension and compression, maximum and minimum stress correspondingly. To analyze the interaction of each one of the two residual stresses and the applied cyclic stresses, let us consider the following two cases.

Case 1   Consider an interaction of a residual compression stress σ_r, with the applied stress σ_{appl} at surface S. See Figure 6.33(a). The stresses are added, producing a nonsymmetric loading cycle with a mean compression stress

$$\sigma_m = \sigma_r \qquad\qquad\qquad [6.23]$$

The maximum combined stress equals the sum

$$\sigma_{max} = \sigma_m + \sigma_{appl,max} \qquad\qquad\qquad [6.24]$$

and may be either tensile or compressive, depending on the magnitude of the mean stress. The minimum combined stress equals the sum of the residual compression stress and the mean stress which is also a compression stress.

Case 2   Consider an interaction of a residual tension stress with the applied stress. See Figure 6.33(b). The stress cycle becomes again nonsymmetrical and the resulting mean stress is a tension stress. The maximum stress, $\sigma_{appl,max}$, is the sum of residual tension stress and maximum applied stress. The presence of residual tension stresses in the surface layer is a detrimental factor because the maximum stresses are increased which, in turn,

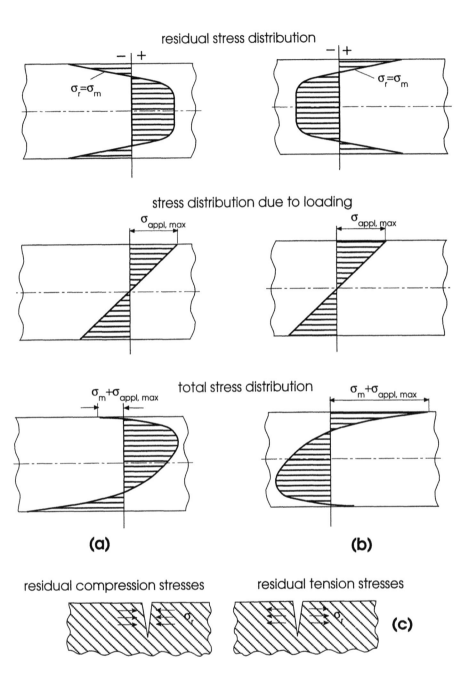

residual stress distribution

stress distribution due to loading

total stress distribution

(a)   (b)

residual compression stresses   residual tension stresses

(c)

Figure 6.33 Stress distribution in a part with residual stresses:
(a) residual compression stresses; (b) residual tension stresses;
(c) effect of residual stresses upon cracks.

accelerates the fatigue failure. On the other hand, the effect of residual compression stresses is a favorable one, slowing down the fatigue failure and extending fatigue life.

Since the fatigue failure usually begins with crack propagation, let us consider the effect of residual stresses upon surface cracks. Tensile stresses accelerate crack growth, while compressive stresses retard the crack propagation. See Figure 6.33(c). In stress concentrations caused by notches, residual compression stresses eliminate the negative effect through redistribution of stresses at the notch root.

In general, when residual compression stresses are present in the surface layer of a part, the fatigue limit increases, and with residual tension stresses present the fatigue limit decreases. This effect is especially pronounced when the tensile and compressive strengths of the metal differ significantly. In parts made of hardened high strength steels, this difference is quite large, causing a particularly strong dependence of the fatigue limit on the residual stresses. In parts made of ductile nonhardened steels, the dependence is reduced. Figure 6.34 presents the relationship between the fatigue limit of alloy steels and residual stresses caused by rolling. According to test data, the fatigue limit can be approximated by the linear function

$$\sigma_{f,r} = \sigma_f - K\sigma_r \qquad\qquad [6.25]$$

where σ_f is the fatigue limit in the absence of residual stresses, while K is an empirical factor which ranges from 0.1 to 0.3 (high values of K are typical for high-strength metals). According to the above equation, residual tension stresses (which have positive sign) reduce the fatigue limit and residual compression stresses (which have a negative sign) increase it. The changes in the fatigue limit may reach in excess of 30 percent.

To take into consideration the temperature factor, the above equation is applicable to machine parts at room temperature. At elevated temperatures, as explained above, a relaxation of residual stresses in the surface layer takes place, accelerating at higher temperatures. At temperatures equal to the recrystallization temperature of the mate-

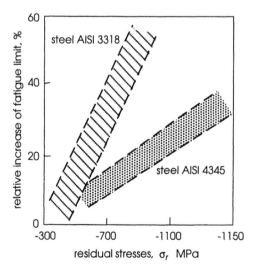

Figure 6.34 Fatigue limit increase versus residual stresses.

rial, the residual stresses are almost completely eliminated. Therefore, at high temperatures, the residual stresses have practically no effect on the fatigue limit.

6.4
PROCESSING METHODS AND FATIGUE: SUMMARY

As explained, the fatigue strength of a machine part is affected by surface integrity, including surface roughness, cold work, residual stresses and structural transformations in the surface layer. The extent to which these factors are effective and the relation between them depend on the processing methods, materials, and working conditions of the part. Thus, when the surface roughness changes from $R_a = 0.16 \, \mu$m to $R_a = 2.5 \, \mu$m, the fatigue limit decreases about 10 to 30 percent. See Reference 3. Residual stresses are significant in parts made from hardened high-strength steels and alloys, while cold work in the surface layer affects mainly parts made from ductile metals, non-hardened steels and alloys. At high temperatures (700 to 900°C) the main factors affecting the fatigue limit are surface roughness and metal structure, since cold work and residual stresses in the surface layer disappear.

Table 6.10 presents the variation in the fatigue limit of the nickel-chromium steel AISI 4345 versus processing. In grinding with subsequent polishing the fatigue limit equals 640 MPa. A complementary lapping reduces the height of irregularities by a factor of six, resulting in a roughness of $R_a = 0.12 \, \mu$m to $0.02 \, \mu$m, thus increasing the fatigue limit by 5 percent without changing the state of the surface layer. Electropolishing following the grinding and polishing removes the layer with surface defects, causing the fatigue limit to rise by about 33 percent while the surface roughness remains unchanged. A rolling after the grinding and polishing produces surface-layer cold work, causing residual compression stresses of about 1600 MPa and increasing the fatigue limit to 900 MPa (40 percent higher than without rolling). Structural defects reduce the fatigue limit by 32 to 48 percent, but appropriate finishing methods may raise the fatigue limit to its initial value and even higher. Table 6.11 shows the effect of processing on the fatigue limit in general.

In summary, the choice of optimal finishing methods is dictated by the following requirements:

(a) a low surface roughness is desirable;

(b) valleys between irregularities should not be sharp, for reduced stress concentration;

(c) the desired direction of processing (traces and irregularities) is parallel to the axis of the processed part;

(d) a surface layer containing defects (decarburization, burns, cracks) should be removed.

(e) in parts operating at temperatures below recrystallization, a strain-hardened surface layer is desirable;

(f) in parts operating at temperatures above recrystallization, surface layer strain-hardening should be avoided;

(g) residual compression stresses in the surface layer are desirable and residual tension stresses should be avoided.

TABLE 6.10

Processing Effects on Fatigue Limit of AISI 4345 Steel (HV 570)

Processing Method	SURFACE LAYER CONDITION		Fatigue Limit (MPa)	Relative to Basic Process (percent)
	R_a (μm)	Decarburization	Residual Stresses (MPa)		
Grinding and polishing (basic process)	0.12	—	+100	640	100
Grinding, polishing and lapping	0.02	—	+100	670	105
Grinding, polishing and electropolishing	0.08	—	+40	850	133
Grinding, polishing and rolling	0.12	—	−1600	900	140
Grinding and polishing	0.10	minor	+480	430	67
Grinding, polishing and lapping	0.02	minor	+100	490	77
Grinding, polishing and electropolishing	0.05	minor	+180	770	120
Grinding, polishing and rolling	0.10	minor	−1300	780	122
Grinding and polishing	0.10	average	+650	330	52
Grinding, polishing and rolling	0.10	average	−1300	680	106
Grinding and polishing	0.10	burns	+280	440	68
Grinding, polishing and rolling	0.10	burns	−1600	780	122

Source: Balter, M.A., see Reference 5.

TABLE 6.11
Effects of Mechanical Processing on Fatigue Performance of Parts

Processing Method	R_a (µm)	Strain Hardening	Residual Stresses		Layer Depth	Fatigue Limit Effect
Turning and planing	0.32–3.2	Present	Tensile	Compressive	0.05–0.15	Variable
Boring	0.32–3.2	Present	Tensile	Compressive	0.03–0.10	Variable
Milling	0.32–5.0	Present	Tensile	Compressive	0.02–0.06	Variable
Broaching	0.63–2.5	Present	Tensile	Compressive	0.05–0.15	Variable
Reaming	0.32–1.25	Present	Tensile	Compressive	0.01–0.03	Variable
Grinding	0.2–1.25	Absent	Tensile	—	0.02–0.04	Decrease
Honing	0.2–0.8	Variable	Tensile	—	0.01–0.02	Decrease
Polishing	0.04–0.32	Variable	Tensile	Compressive	0.01–0.02	Increase
Lapping	0.01–0.16	Variable	Tensile	Compressive	0.01–0.02	Increase
Superfinish	0.02–0.16	Variable	Tensile	Compressive	0.005–0.01	Increase
Vibroabrasive processing	0.3–2.5	Present	Tensile	Compressive	0.01–0.02	Increase
Electropolishing	0.15–0.3	Absent	—	—	—	Increase
Rolling	0.08–0.63	Present	—	Compressive	0.1–1.0	Increase*
Burnishing	0.04–0.63	Present	—	Compressive	0.1–0.3	Increase*
Tumbling	0.32–0.63	Present	—	Compressive	0.1–0.3	Increase*
Shot peening	1.25–5.0	Present	—	Compressive	0.05–0.3	Increase*
Microshot peening	0.16–0.65	Present	—	Compressive	0.01–0.05	Increase*

*significant increase

REFERENCES

1. "Machining Data Handbook". 1980. Vol. 2, 3rd ed. Cincinnati: Metcut Research Associates Inc.
2. Neuber, H. 1958. *Kerbspannungslehre.* Berlin: Springer Verlag.
3. Sulima, A.M., and M.I. Evstigneev. 1974. *Surface Integrity and Fatigue of Parts from High-Temperature and Titanium Alloys.* Moscow: Mashinostroyenye. (In Russian.)
4. Shnaider, Y.G. 1967. *Cold Working of Metals.* Moscow: Mashinostroyenye. (In Russian.)
5. Balter, M.A. 1978. *Hardening of Machine Parts.* Moscow: Mashinostroyenye. (In Russian.)
6. Khvorostukhin, L.A. 1974. *Improvement of Bearing Capacity of Machine Parts by Surface Strain Hardening.* Moscow: Mashinostroyenye. (In Russian.)
7. Kishkin, S.G., and A.M. Sulima. 1956. "Cold-Work Influence on Mechanical Properties of High-Temperature Resistant Alloys," *Proceedings of Moscow Aviation Institute,* 71. (In Russian.)

7

FATIGUE LIFE IMPROVEMENT

In Chapter 6 it was shown that the mechanical processing methods have a profound effect on the fatigue strength of a machine part. The surface layer may be either strengthened or weakened, depending on the processing conditions. The optimal surface integrity may be achieved by (a) a proper choice of conditions when applying conventional processing methods (b) special strengthening methods or (c) selecting an optimal sequence of selected processing operations.

The integrity of surface is determined during the stage of finishing. The processing operations preceding it affect the surface layer as well, however, and dictate the kind of finishing method that is to follow. If the preceding operations leave a positive effect on the surface (such as strain hardening and residual compression stresses), the finishing operations are chosen to retain this effect. To the contrary, with negative characteristics present (such as decarburized surface layer or residual tension stresses), the finishing operations ought to remove them. Among the finishing operations, producing surface layer properties that extend fatigue life, the most common are cold-working methods such as shot peening, rolling, and burnishing.

7.1
SHOT PEENING

The most widespread cold-work method used for surface layer strain hardening today is shot peening. In this method, streams of metallic shot (particles) are directed at the processed part at velocities reaching 100 m/sec. The particles impinge upon the processed surface, causing indentations and a uniform strain hardening of the surface layer. The main parameters of shot peening are particle dimensions (0.1 to 5 mm), material of particles (steel, cast iron, glass, quartz), particle velocity (1 to 100 m/sec), shot flow rate (kg/min), and processing time (minutes).

The particle dimensions, material properties and velocity are the parameters that determine the kinetic energy of a particle

$$E = \frac{\pi\gamma}{12g} D^3 V^2 \qquad\qquad [7.1]$$

where D is the particle diameter, γ is the specific weight, V is the velocity, and g denotes the gravitational constant. Kinetic energy controls the depth of indentation, the thickness of hardened layer, the rate of hardening and the magnitude of residual stresses. The other two parameters, the flow rate of particles and processing time, determine the continuity and uniformity of the surface and the rate of hardening.

There are two ways to implement shot peening: the conventional application known as the "dry" method and the "wet" method. In the wet method, the particles hit the surface through an oil film. Because of the softer impact, the processed surface layer is substantially thinner than that of the dry method and the stresses are more uniform.

7.1.1 Shot peening devices

Most shot peening machines impart the kinetic energy upon particles either by the centrifugal force or by a jet of air or liquid. Figure 7.1 shows two centrifugal shot peening machines—"dry" and "wet." Let us consider the performance of a dry machine first. See Figure 7.1(a). The shot is placed into hopper 1, from there it is lifted by elevator 2 into hopper 3. After entering rotating wheel 4, the shot is ejected, impinging upon the processed surface S. We describe the wet centrifugal shot peening machine by illustrating, in Figure 7.1(b), a machine designed for cold-working of turbine blades. A number of blades B are fixed in rotating pallets 1. Working mixture 2, comprised of particles and oil, is lifted from the bottom of tank 3 by a centrifugal action of propeller 4 and ejected onto the blades. The shot hits the blades through an oil film.

Jet shot peening machines cause particle acceleration by either pneumatic (dry) or hydraulic (wet) means. See Figure 7.2. Part (a) of the figure shows a dry shot peening device. A jet of air under pressure from ejector 1 delivers the shot onto the processed surface S. A wet machine is shown in Figure 7.2(b). It comprises chamber 1, divided by a wire net 2 into two parts, 3 and 4. The bottom of 4 is filled with liquid, usually machine oil. Shot 5 is placed in part 3 and set into pulsating motion. An ejector nozzle 6 at the bottom of part 3 generates a jet of liquid under pressure, which picks up the shot and directs it at processed surface S. In general, hydraulic ejection machines are

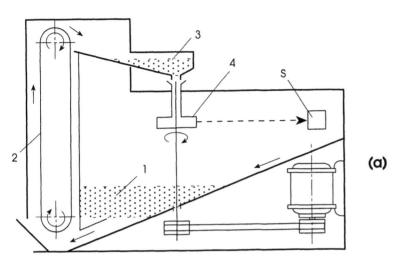

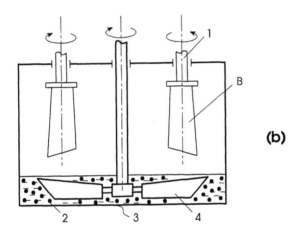

Figure 7.1 Centrifugal shot-peening machines: (a) dry shot-peening machine: (b) wet shot-peening machine. S denotes shot-peened surface, B denotes shot-peened blade.

used for processing more complicated parts where more accurate results are required. Using machines similar to that shown in Figure 7.2(b), with microballs of 0.05–0.3 mm in diameter, intricate parts can be processed with thin edges and small fillet radii and without deformation, achieving surface roughness of $R_a = 0.16$–0.63 μm with very thin strain-hardened layers.

Figure 7.3 shows two pneumatic shot peening devices for local cold working of large parts. The first device, see Figure 7.3(a), comprises a conical chamber 1 with particles 2, located next to the processed surface S and a container of compressed air 3. The particles are set in motion by a jet of air 4, reaching velocities of 10 to 15 m/sec. The

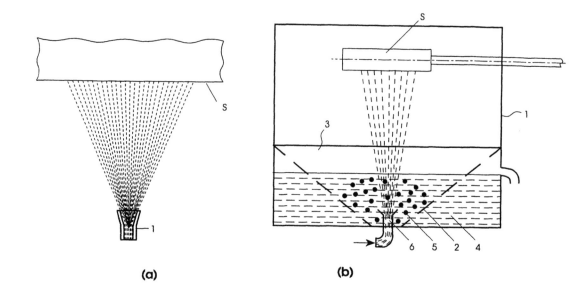

Figure 7.2 Jet-ejection shot-peening machines: (a) dry
shot-peening machine; (b) wet shot-peening machine.

particles, after hitting the processed surface, roll down to the bottom of conical chamber
1 where they are picked up by the air jet again. The device may be moved along surface
S, to process the entire area. Changing the air pressure, feed, particle size and number
of particles, a wide range of surface parameters can be achieved. Figure 7.3(b) shows
a second shot peening device for local cold working of holes. The processed surface S
is impinged by particles 1 that get the kinetic energy from high-speed impeller 2. The
device can be moved axially along the hole, strain-hardening the whole surface. Shot
peening devices used for limited areas have an advantage of being small in size and
requiring a lesser quantity of particles.

7.1.2 Surface integrity

The surface texture resulting from shot peening is an aggregate of a large number
of superimposed indentations—traces of particle impingements. The irregularities are
measured by the depth of indentations and the height of plastic burrs at the edges. The
valleys have a smooth curvature, created by the impact of round particles. The valley
radius r_v matches the radius of particles (usually 0.2–1 mm) and is much larger than
the radius r_v which results from cutting and abrasive processing. The surface roughness
caused by shot peening has a favorable effect upon stress distribution. Even for large
R_a, the stress concentration in valleys after shot peening is low.

The rate of hardening in shot peening is 20 to 40 percent, depending on the
processed material and the kinetic energy of particles. The residual stresses in the
surface layer are always compressive, reaching 400–800 MPa. The combination of
residual compression stresses and round valleys is of special importance to machine parts

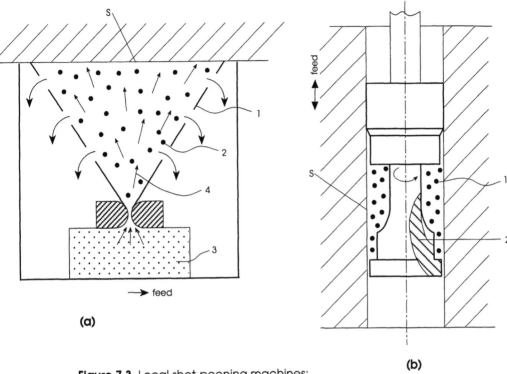

Figure 7.3 Local shot-peening machines:
(a) for flat surfaces; (b) for holes.

which operate under cyclic loading such as springs, gears, shafting, and others. Shot peening increases the fatigue life of spiral springs by a factor ranging from 8 to 14, of leaf springs in trucks by a factor of 14, and of coarse gears by a factor of 6 to 12.

The dry and wet methods produce different surface integrities. In the dry method the surface roughness is $R_a = 2.5–5.0 \ \mu$m and the depth of strain-hardened layer is up to 1.5 mm. The wet method gives a lesser roughness of $R_a = 0.16–1.25 \ \mu$m, with a strain-hardened layer depth of up to $0.1 \ \mu$m. The wet method is therefore a preferable finishing operation.

Let us now consider compressor blades of a turbojet engine and the effect of processing on the fatigue limit, as presented comparatively in Table 7.1. The data shown are for blades produced with the required precision by the initial processes of milling, grinding and polishing followed, in the latter stage, by a partial removal of grinding defects, including residual tension stresses.[1] At this stage the blades have a low surface roughness ($R_a = 0.16–0.32$ mm), a strain-hardened surface layer of about 0.02 mm thick and fatigue limit of 240 MPa. In the surface layer are small residual compression stresses, below which the stresses are tensile. As can be seen from the table, the finishing operation of etching increased the fatigue limit by 13 percent, while rolling, produced a strain-hardened surface layer 0.05 mm thick with residual compression stresses of $\sigma_r = -400$ MPa. The increase in fatigue limit after rolling was 15 percent. Compare this with the data after the dry shot peening was performed and caused strain hardening with residual compression stresses of $\sigma_r = -550$ MPa. Note that despite the

TABLE 7.1
Effect of Processing on Fatigue Limit of Machine Parts

Processes	SURFACE INTEGRITY PARAMETERS				FATIGUE LIMIT	
	R_a (mm)	H_μ (MPa)	h_s (mm)	Residual Stresses (MPa)	(MPa)	Ratio to basic processes (percent)
COMPRESSOR BLADE, TITANIUM ALLOY, $R_c = 27$-36						
Basic: milling, grinding and polishing	0.16–0.32	382	0.02	−150	240	100
Basic processes and etching	0.32–0.63	345	—	—	270	113
Basic processes and rolling	0.32–0.63	420	0.05	−400	275	115
Basic processes and "dry" shot peening	2.5–5.0	440	0.08	−550	310	129
Basic processes and "wet" shot peening	0.16–0.63	450	0.05	−490	340	142
COMPRESSOR BLADE, HIGH TEMPERATURE RESISTANT STEEL, BHN = 311–363						
Basic: milling, grinding and polishing	0.32–0.63	360	0.02	−150	300	100
Basic processes and "wet" shot peening	0.32–0.63	445	0.08	−380	400	133
GEAR, STEEL 3312, CARBURIZED, $R_c = 60$-62						
Basic: gear grinding	0.63–1.25	710	—	+300	60	100
Basic process and "wet" shot peening	0.63–1.25	850	0.1	−900	85	144
CONNECTING ROD, STEEL 3318, $R_c = 38$-40						
Basic processes and die forging	3.2–6.3	—	—	—	57	100
Basic processes and "wet" shot peening	1.25–2.5	—	—	—	75	132
BOLT, ALLOY STEEL 3130						
Without cold working	—	—	—	—	260	100
Shot peened by microballs ($d = 0.16$–0.2mm)	—	—	—	—	440	170

Source: Petrosov, V.V., see Reference 1. Yatsenko, V. K., et al., see Reference 6.

surface roughness deterioration to the value $R_a = 2.5–5.0$ mm, the fatigue limit was raised to 310 MPa, a 29 percent increase. A wet shot peening produced even better results with the fatigue limit raised to 340 MPa while the surface roughness remained unchanged. This is an increase of 42 percent. It follows that the shot peening operation has a considerable effect in extending the fatigue life of a machine part.

7.2
ROLLING

Next to shot peening, rolling is the most popular cold-working operation. Producing low surface roughness, rolling processing increases the fatigue limit of a machine part and is applicable to external as well as internal surfaces. Contrary to shot peening, there is no need for special equipment, since rolling may be performed with standard machining devices such as a lathe. The process of rolling involves plastic deformation of the processed surface by a rolling tool—either a roller or a ball. The rolling is illustrated in Figure 7.4. Part (a) shows typical rolling operations, while part (b) presents a number of different rollers commonly used. The rollers are made from hardened tool

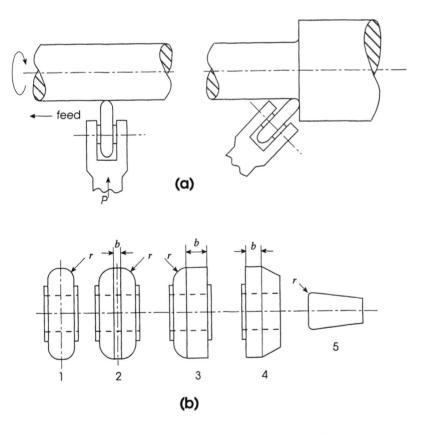

Figure 7.4 Rolling devices: (a) rolling of axisymmetric surfaces: (b) profile rollers.

steels and cemented carbide alloys. Their working surface is thoroughly ground and polished. During processing, the roller reduces surface irregularities and strain hardens the surface layer, creating residual compression stresses.

7.2.1 Parameters of rolling

The main factors in a rolling operation are the rolling force and feed. The rolling force determines the deformation of the surface layer, affecting such parameters as surface roughness, strain hardening and residual stresses. The force applied must produce sufficient strain hardening of the surface layer without damaging the layer. The magnitude of the rolling force depends on mechanical properties of the processed metal, the dimensions of the processed part and tool, as well as other parameters. While the rolling force is, usually, determined experimentally, it may also be estimated using the following expressions:

for operations controlled by a roller

$$P = \frac{320Ddb}{E(D+d)}\sigma_{yp}^2 \qquad [7.2]$$

for operations controlled by a ball

$$P = \frac{4 \cdot 10^4 d^2}{E^2}\sigma_{yp}^3 \qquad [7.3]$$

where P denotes the rolling force, D is the diameter of the workpiece, d is the diameter of the roller or ball, and b is the width of the cylindrical part of the roller. E is the Young's modulus of the processed material and σ_{yp} is its yield point.

In rolling the second important factor is feed, chosen to assure a uniform processing of the surface. Its size should be smaller than the width of the contact area between the tool and the workpiece. The ratio between width of the contact area and the feed usually equals

$$\frac{a}{f} = 3 \text{ to } 6 \qquad [7.4]$$

where a is the width of the contact area and f is the feed. The feed has little effect on strain hardening but it affects strongly the surface roughness. See Section 6.1. Other parameters (such as velocity, number of passes, type of lubricant used) all have a lesser effect on rolling operation.

7.2.2 Surface integrity

As in other processes, the surface integrity of the rolled surface also depends on the material of the processed part, initial state of the processed surface, and the processing conditions. Under optimal conditions, the height of irregularities is reduced by this process by a factor of 4 to 10. The obtained surface roughness, R_a, ranges from 0.16 to 1.25 μm. The depth of strain-hardened layer varies from 0.2 to 25 mm. In rolling the surface hardness increases by 10 to 50 percent, accompanied by corresponding residual compression stresses.

The surface integrity is sensitive to instability in the rolling force. Using tools with elastic elements (such as springs, hydraulic dampers, and others) guards the stability. These elastic elements eliminate knocking of the rotating workpiece. In the following discussion we shall consider the effect of rolling upon machine parts with different shapes.

Rolling of shafts

Rolling is one of the most common operations among the cold-working processes of axisymmetric bodies such as shafts. For these parts, the rolling is more convenient than shot peening.

Let us consider rolling of a forging hammer as illustrated in Figure 7.5. The shank, after fine turning with surface roughness of $R_a = 5-10\ \mu$m, is vulnerable to breakage under fatigue loading. The fatigue fracture usually occurs at the base of the conical section. A number of forging hammers were processed by rolling in order to investigate its effect on the fatigue life. The danger zones were rolled on a lathe with a roller tool, see figure. The rolling conditions were as follows:

diameter of roller	105 mm
profile radius of roller	7 mm
rolling force	35 kN
feed	0.4 mm/rev
velocity	30 m/min.

As a result of rolling, the parameters of the processed units were as follows: the surface roughness improved to $R_a = 0.63-1.25\ \mu$m, the fatigue limit was raised by 40 to 60 percent and the fatigue life increased by a factor of 2.3 to 2.7.

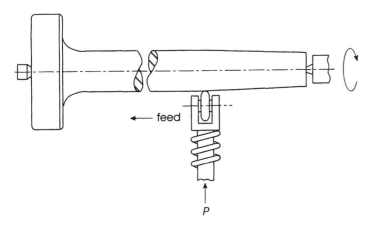

Figure 7.5 Rolling of a shank of die-forging hammer.

Rolling of fillets

Fillets are the locations of stress concentration in shafts and other axisymmetric bodies and therefore deserve a special attention in cold working by rolling. Figure 7.6. shows several cases of fillet rolling. The simplest case uses a roller (or a ball) with a radius equal to that of fillet. The roller (or a ball) is set at an angle of 45 degrees to the processed part. See Figures 7.6(a) and (b). This method is effective for small radii only, because the required rolling force for larger radii may be off limits for the processed part and equipment. To accommodate fillets with a large radius, one uses a combination of small-radius rollers with either longitudinal or circular feed. See Figures 7.6(c) and (d). The feed of such rollers, however, requires special devices. In addition, there is another method for large fillet radii, using only rollers, specially shaped and set at 45 degrees, as shown in Figures 7.6 (e) and (f).

To illustrate a special case, Figure 7.7 shows fillet rolling in a crankshaft. The rolling operation is accomplished by rotating the crankshaft in a special rolling device. The device comprises rollers 1 which are supported by wheel 2 in housing 3. Supporting rollers 4 are provided to prevent deformation of the crankshaft. The rolling force P is controlled by a special hydraulic device. As a result of rolling, the fatigue limit of the crankshaft rose by 55 to 75 percent.

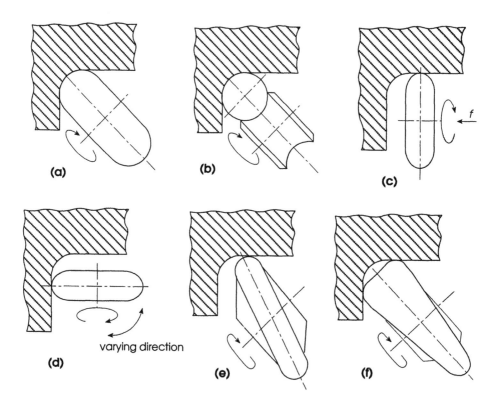

Figure 7.6 Rolling of fillets.

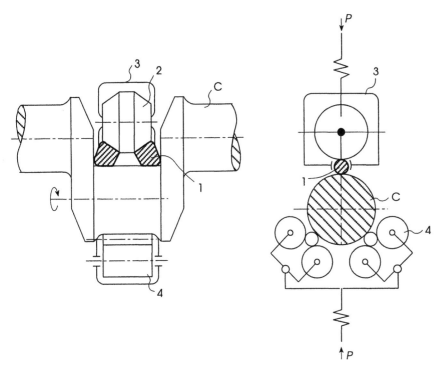

Figure 7.7 Fillet rolling of a crankshaft with profile rollers.

Table 7.2 shows test results of different shafts, where the fatigue limit rose by 30 to 80 percent after rolling.[2,3] Table 7.3 presents a comparison of fatigue limits of smooth shafts without rolling and multidiameter shafts with fillets subject to rolling. It follows from the data that rolling fully eliminates the effect of the stress concentration in the fillets.

Rolling of screw threads

The methods of thread forming, cutting, rolling, or both, affect the surface integrity variously. The shape of the threads is the source of high stress concentration, the magnitude of which depends on the radius at the root of the thread. Thread rolling has a profound effect on fatigue strength in cyclic loading. There are two applications of thread rolling: first, original forming, and second, superficial rolling of roots after cutting and grinding to improve surface integrity. In the latter application, the roller used to improve the cut thread has profile angle 10 to 20 degrees smaller than the thread angle and a radius slightly smaller than the root radius. The roller comes in contact with the root of the thread only.

Table 7.4 shows the effect of different thread forming methods on the fatigue limit.[4] The data indicate a significant increase of fatigue limit due to rolling, up to 60 percent.

One highly efficient method applicable in cold working of threads is a combination of rolling and roller peening under dynamic forces. It pertains to threads with a pitch exceeding 8 mm. Its effect depends on the deformation rate of the processed metal. Table 7.5 lists values of the fatigue limit for different rates of deformation. The data

TABLE 7.2
Effect of Rolling on Fatigue Limit of Multidiameter Shafts

SHAFT DIMENSION			ROLLER		FATIGUE LIMIT		
Small Diameter D (mm)	Fillet radius r (mm)	r/D (ratio)	Profile Radius (mm)	Force P (kN)	Unrolled (MPa)	Rolled (MPa)	Increase (percent)
12	0.8	0.066	0.8	0.15	18	19	5
12	0.8	0.066	0.8	0.45	18	25	33
14	0.8	0.057	0.75	2.5	18	29	17
14	0.8	0.057	0.75	5.0	18	29	61
15	2	0.133	2	3.5	28	36.4	30
17	1	0.06	1	7.0	20.8	35.2	68
18	0.76	0.042	0.76	2.5	15	23.4	57
30	2	0.066	2	1.5	14	18	29
30	2	0.066	2	2.5	14	21	50
30	2	0.066	2	4.5	14	26	86
45	2	0.045	1.97	4.0	12	21	75
133	7.15	0.053	4.4	6.8	13	16.8	30
180	8	0.045	7.93	65	13	18.5	42

Sources: "ASME Handbook, Metals Engineering Design," see Reference 2; Braslavsky, V.M., see Reference 3.

indicate that even at deformation rates of tenths of a percent, the fatigue limits of the threaded parts increase by 80 to 180 percent. The combined rolling-peening method is used in cold working of large parts such as shafts, axles, and others.

Gears and splines

In most cases, fatigue failure of gears and splines in transmission systems is caused by stress concentration at the roots of the teeth. Additional stress concentration comes from irregularities in the surface.

TABLE 7.3
Fatigue Limits of Shafts After Processing

Material	Shape	Processing	r/D	Fatigue Limit (MPa)
Steel AISI 4330	Smooth	Turning	—	290
	Stepped with fillet	Turning	0.022	88
	Stepped with fillet	Fillet rolling	0.022	288
Steel AISI 1045	Smooth	Turning	—	191
	Stepped with fillet	Turning	0.022	76
	Stepped with fillet	Fillet rolling	0.022	190

TABLE 7.4
Fatigue Limits of Thread After Processing

Material	Thread Processing M12 × 1.5	Fatigue Limit (MPa)
Steel AISI 4340	Grinding	90
	Grinding and rolling of roots	145
	Thread rolling	135
	Thread rolling and rolling of roots	160
Steel AISI 1045	Turning and grinding	70
	Grinding and rolling of roots	100
	Thread rolling	95
	Thread rolling and rolling of roots	112

Source: Rikovsky, B. P., V. A. Smirnov and G. M. Shetinin, see Reference 4.

Rolling operation is an effective means for strengthening gear teeth. The method is suitable for strain hardening of medium-sized teeth with a modul of $m = 7-10$ mm. The constraints are that in smaller-sized teeth the space between is too small to accommodate a roller, and for larger teeth the needed forces are too large. The forces can be reduced using rollers with a smaller radius. Table 7.6 presents comparative test data showing that rolling operation raises the fatigue limit of gear teeth by 18 to 26 percent.

Rolling of splines has been successfully applied to increase fatigue limits of spline shafts. Bending tests of a spline shaft $D = 230$ mm in diameter, made of alloy steel AISI 4334, showed that rolling increased the fatigue life from 52000 to 160000 cycles. See Reference 3.

TABLE 7.5
Fatigue Limits of Roller-Peened Buttress Thread

Deformation rate (Δd*/d*) 100%	FATIGUE LIMIT			
	STEEL AISI 1040		STEEL AISI 4335	
	(MPa)	(percent)	(MPa)	(percent)
0	115	100	90	100
0.1	125	109	125	139
0.25	150	130	165	183
0.50	175	152	198	220
0.75	200	174	230	255
1.00	216	188	250	278

Source: Braslavsky, V. M., see Reference 3.

d*—root diameter of the thread

TABLE 7.6
Fatigue Limits of Spur Gears

	FATIGUE LIMIT		
Material	No Cold Working (MPa)	Rolled (MPa)	Increase (percent)
Alloy steel AISI 4137 hardened	300	380	26
Carbon steel AISI 1055 hardened	320	400	25
Alloy steel AISI 4334 non-hardened	340	410	20
Carbon steel AISI 4334 hardened	620	730	18

Source: Braslavsky, V. M., see Reference 3.

7.3
BURNISHING

Cold-working methods, such as shot peening, rolling and others, are applicable to processing of machine parts with hardness not exceeding $R_c = 40-50$. This is because the hardness of tools (such as metal shot or rollers) is usually $R_c = 60-62$. Many machine parts, however, are manufactured from high strength steels and alloys having a higher hardness. For these machine parts the applicable cold-working processing is burnishing.

The burnishing process uses a sliding tool—the burnisher. The tool is made of super-hard material with a rounded working surface, radius of $0.5-4$ mm, which is set on a standard metalworking machine. Burnishers are manufactured from natural diamonds and from synthetic polycrystal diamonds. See Figure 7.8 where part (a) shows the burnishing of a cylindrical surface on a lathe, while part (b) presents the burnisher. During the operation the surface layer is strain hardened and residual compression stresses are generated, while the previous surface roughness is smoothed out.

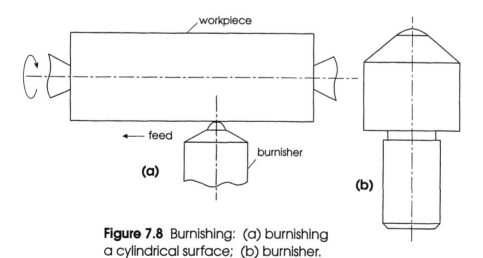

Figure 7.8 Burnishing: (a) burnishing a cylindrical surface; (b) burnisher.

Burnishing has the following features when compared to other cold-working methods:

(a) The superhard materials can be used to burnish any metals, soft or hardened (up to R_c 60 to 64), susceptible to strain hardening, and those that cannot be processed by other cold-working methods.

(b) Due to a small radius of the sliding element of the burnisher, the contact area with a workpiece is small (less than 0.1–0.2 mm^2). This causes high contact pressure and therefore the required burnishing forces are small (50–250 N). These small forces allow burnishing of thin elastic parts without altering their geometry. The process is slow, however, because of the tool's small size.

(c) Sliding friction between the tool and workpiece improves surface integrity which simplifies the design of the burnisher and reduces the dimensions.

The burnished surface is characterized by low roughness height ($R_a = 0.04$–$0.32 \,\mu$m), smooth irregularities (rounded valleys and peaks) high strain hardening (20 to 50 percent) and strain-hardened depth of 0.1–0.4 mm. The surface contains large residual compression stresses, therefore it offers considerable fatigue life improvement.

Processing parameters

The main processing parameters of burnishing are force, feed and velocity. Force being the most important, its magnitude determines the intensity of strain hardening and affects the surface integrity more than other parameters. The usual values of burnishing force are in the range between 100 and 300 N. The burnishing force is directly proportional to the surface hardness, the thickness of strain-hardened layer, and the compressive residual stresses. It is inversely proportional to the surface roughness. As the force increases and reaches a value of about 200–300 N, the surface roughness and hardness stabilize. Above that range, the surface roughness slightly increases, while the surface hardness decreases. The thickness of strain-hardened layer and the residual stresses increase proportionally to the applied force.

The second most important parameter is feed which affects the surface roughness in direct proportion. Strain hardening and residual stresses are affected by feed to a lesser extent. The feed in burnishing usually ranges from 0.02 to 0.08 mm/rev.

The burnishing velocity is not included among parameters affecting surface integrity because of its negligible effect. It is a factor to consider for tool wear and surface defects which are caused by heating, such as softening or burns. Usually the burnishing velocity is in the range of 100 to 150 m/min.

Fatigue life improvement

Let us consider the results from fatigue tests performed on a number of carbon and alloy steels to determine the effect of burnishing on fatigue strength. The tests were carried out in rotating bending at 2,000 revs/min, with asymmetrical cyclic loading. The specimens used had a ground, polished or burnished surface with a roughness of $R_a = 0.08$–$0.16 \,\mu$m. The results are presented in Figure 7.9 which compares the fatigue strengths of burnished and ground specimens. Table 7.7 compares the fatigue limits of burnished and polished specimens.[5] The results indicate that burnishing raises the fatigue limit up to about 60 percent. A fracture analysis of the broken specimens shows

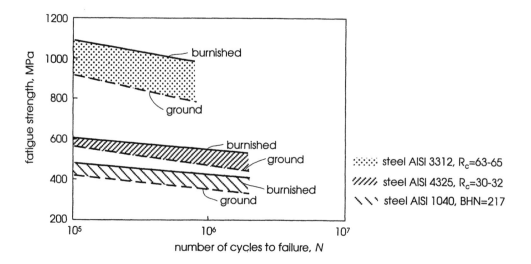

Figure 7.9 Fatigue limit after grinding and burnishing.
(Data from Torbilo, V. M., see Reference 5.)

that the fatigue crack initiation starts at a depth of 0.1–0.4 mm which is below the strain-hardened layer.

Fatigue strength in corrosive media   The investigation of burnishing effects on fatigue strength of specimens in corrosive media was performed at different cyclic lives. Table 7.8 presents the test results which confirm that corrosion causes a considerable reduction of fatigue strength. The harmful effect of corrosion grows in direct proportion to the cyclic life. After prolonged tests (up to 10^8 cycles) it was concluded

TABLE 7.7
Fatigue Test Results in Air

Material	Hardness	FATIGUE LIMIT		
		Polished (MPa)	Burnished (MPa)	Increase (percent)
Steel AISI S2100	$R_c = 64-65$	1040	1180	14
Steel AISI 3312	$R_c = 63-65$	800	1000	25
Tool steel DIN 105WCr6*	$R_c = 65$	610	720	18
Alloy Cr-Ni-Mo steel	$R_c = 30-31$	500	720	43
High temperature resistant alloy steel	BHN = 250–265	560	670	19
Steel AISI 4335	$R_c = 30-32$	520	590	14
Stainless steel AISI 321	BHN = 175	250	410	63
Carbon steel AISI 1045	BHN = 217	350	420	20
Carbon steel AISI 1050	BHN = 150	260	320	23

Source: Yatsenko, V. K., et al., see Reference 6.

*German standard

TABLE 7.8
Fatigue Limits of Polished and Burnished Specimen in Corrosive Media

			FATIGUE STRENGTH (MPA)			
Material	Hardness	Processing Method	$N = 10^7$	$N = 5 \times 10^7$	$N = 10^8$	Fatigue Limit in air (MPa)
CORROSIVE MEDIUM: WATER						
Steel AISI 4335	BHN = 320	Polishing	150	100	90	520
		Burnishing	450	320	230	290
Steel Cr-Mn-Si-N	BHN = 320	Grinding	80	—	—	620
		Burnishing	760	—	—	780
Steel AISI 4340	BHN = 320	Polishing	300	240	225	490
		Burnishing	450	430	425	610
CORROSIVE MEDIUM: 4% NACL WATER SOLUTION						
High temperature resistant alloy	BHN = 320	Grinding	340	—	—	500
		Burnishing	460	—	—	650

that burnishing increases the fatigue limit of a specimen in corrosive medium by a factor of about 2. Increased fatigue strength of a burnished specimen in a corrosive medium can be explained by favorable surface properties, such as low and rounded roughness, significant strain hardening, structural uniformity of the surface layer, and residual compressive stresses. The latter two delay the crack initiation and propagation. The residual compression stresses reduce maximum stresses in a loading cycle and prevent a penetration of corrosive medium into the crack opening.

Stress concentration Tests were performed to investigate the influence of burnishing on fatigue limit in the presence of stress concentration. The results of the tests, given in Table 7.9, indicate that burnishing offsets the adverse effect of stress concentration, as compared with turning, grinding, and polishing.[6] After burnishing the fatigue limit of a specimen with fillet, having a radius $r = 1$ mm, reached the value of 500 MPa. (After grinding the value was 340 MPa and after polishing it was 370 MPa.) Burnishing eliminates the effect of stress concentration in grooves where depth does not exceed the thickness of strain-hardened layer containing residual compression stresses. This applies to grooves whose depth is about 0.25 mm or less.

Fatigue strength at high temperatures Machine parts made of high-strength alloys and working at high temperatures (as, for instance, parts of turbojet engines) are subject to thermal fatigue. In such parts a relaxation of strain hardening takes place in the surface layer which changes the fatigue strength characteristics. To determine the burnishing effect upon relaxation, fatigue tests with burnished specimens were performed at high temperatures. The test results are given in Table 7.10. The tests show that strain hardening and high fatigue strength are maintained up to the beginning of metal recrystallization (about 450 to 750°C for steels and nickel-base alloys). The data

TABLE 7.9
Fatigue Limits of Specimen with Stress Concentration
(Cr-Ni Alloy Steel)

	FATIGUE LIMIT			
Stress Concentrators	*Ground (MPa)*	*Polished (MPa)*	*Turned (MPa)*	*Burnished (MPa)*
Fillet, *r* = 10mm	520	550		690
Fillet, *r* = 10mm	480	500		620
Fillet, *r* = 10mm	440	480		580
Fillet, *r* = 10mm	340	370		500
Groove, *r* = 0.25mm			230	420
Groove, *r* = 0.50mm			240	490

Source: Yatsenko, V. K., et al., see Reference 6

TABLE 7.10
Fatigue Limits at High Temperatures

		FATIGUE LIMIT		
Material	*Finishing Method*	*Test Temperature (°C)*	*(MPa)*	*(percent)*
High temperature resistant nickel alloy	Grinding	20	315	100
		300	305	97
		600	290	92
		780	290	92
High temperature resistant nickel alloy	Burnishing	20	505	160
		300	465	146
		600	405	128
		780	350	111
Stainless Steel 321	Polishing	250	201	100
		450	192	96
		600	156	78
Stainless Steel 321	Burnishing	250	365	182
		450	320	160
		600	191	95

Source: Yatsenko, V. K., et al., see Reference 6.

demonstrate that burnishing raises significantly the fatigue limit at high temperatures as compared to grinding and polishing. Even though, at higher temperatures, the stress limit decreases in all tested cases (burnished, ground and polished specimens), an advantage of burnished specimens is noted up to the temperatures of 600 to 700°C.

In summary, burnishing is being applied as a cold-working finishing operation to a wide range of machine parts, such as shafts, bars, pistons, sleeves, cylinders, and others. Usually these parts are manufactured from heat-treated carbon and alloy steels and are ground or finely turned before burnishing. Burnishing then ensures high surface integrity parameters and high endurance, reliability and long fatigue life. Table 7.11 presents test data which confirms the advantage of burnishing as a finishing process after fine turning and grinding.

TABLE 7.11

Surface Integrity Parameters and Fatigue Limits of Gas-Turbine Engine Shafts (High-temperature Resistant Alloy Steel BHN = 300–330)

| | PARAMETERS | | | | | FATIGUE LIMIT | |
Processing	R_a (μm)	H_μ (MPa)	U_s (percent)	h_s (mm)	Residual Stresses (MPa)	(MPa)	(percent)
Fine turning	2.0	3830	11	0.09	−200	440	100
Grinding	0.85	3600	5	0.04	+200	520	118
Grinding and polishing	0.25	3630	6	0.03	+100	560	127
Fine turning and burnishing	0.4	4260	23	0.19	−900	650	148
Grinding and burnishing	0.25	4090	19	0.20	−1000	690	157

Source: Yatsenko, V. K., et al., see Reference 6.

REFERENCES

1. Petrosov, V.V. 1977. *Shot Peening of Machine Parts and Tooling.* Moscow: Machinostroyenye. (In Russian.)
2. "ASME Handbook, Metals Engineering Design." 1965. Ed. Horger, O.J. New York: McGraw-Hill.
3. Braslavsky, V.M. 1975. *Rolling of Large Machine Parts.* Moscow: Machinostroyenye. (In Russian.)
4. Rikovsky, B.P., V.A. Smirnov and G.M. Shetinin. 1985. *Local Cold-Working of Machine Parts.* Moscow: Machinostroyenye. (In Russian.)
5. Torbilo, V.M. 1972. *Diamond Burnishing.* Moscow: Machinostroyenye. (In Russian.)
6. Yatsenko, V.K., G.Z. Zaitsev, V.F. Prichenko and L.I. Ivchenko. 1985. *Improvement of Bearing Capacity of Machine Parts.* Moscow: Machinostroyenye. (In Russian.)

8

DIAGNOSIS OF FATIGUE CASES

To illustrate the practical application of the theories used in fatigue analysis, three actual fatigue cases have been chosen for investigation. The cases come from three different industries: aircraft, military and oil refineries. The cases were chosen for their relevance and diversity of problems: Case 8.1, Fatigue of a Fuse Pin, deals with a sudden fracture due to the damaged surface in a part; Case 8.2, Crack Propagation in a Breech Block, investigates crack progression and cessation; and Case 8.3, Thermal Fatigue of a Coke Drum, is concerned with the fatigue caused by thermal stresses. In general, the reason for investigating fatigue cases after the fact is to ascertain with a reasonable accuracy the causes for unexpected failures and provide a remedy, if possible, with improved design. One of the cases presented here brings out the fact that a failure occurred in spite of a failure proof design and it appears that different inspection procedures could have prolonged the fatigue life. In another particular case, our review shows that what would be generally viewed as a danger signal, crack occurrence, does not affect the operational life of the machine part. The last case shows the cause for failure is such that it can not be remedied by changes in design and a timely replacement or repair must be considered.

The applicable theories used in the analyses include the stress and strain methods and the crack propagation theory. The analyses are done with the help of finite element (FE) programs adapted to nonlinear problems, taking into consideration elastic-plastic behavior of the material. The analyses are to be taken as illustrative examples only, since the data comes to us from outside sources. Our conclusions are in accordance with limited understanding of the facts and conditions in each case described and more accurate results might be achieved if more precise information were available.

8.1
FATIGUE OF FUSE PIN IN AN AIRCRAFT

The purpose of this investigation is to analyze an unexpected fracture, due to fatigue, of a fuse pin in pylon mounting which caused an aircraft crash. The analysis finds that the fatigue design of the critical part was not at fault: the provided magnitude of life expectancy was to assure a long uninterrupted performance. The reason for a sudden failure seems to lie in another factor with as strong effects on fatigue life, namely, the surface condition of a part. Fatigue failure begins in most cases at the surface which bears the greatest load and is exposed to environmental effects. As it appears from our analysis, a local surface defect inside the fuse pin may have been the source of a fatigue crack which propagated and caused the critical fracture.

8.1.1 The problem

In October 1992 a fully loaded cargo aircraft Boeing 747 of the El Al Airlines, after taking off from Shiphol Airport in Holland, lost two of its four turbofan Pratt and Whitney JT9D-7J engines and crashed. According to the findings of a formal investigation,[1] the accident occurred when one pylon and the attached engine separated from the wing, hitting the adjacent engine and also tearing that one off. At the time, the aircraft accumulated 45,746 flight hours corresponding to 10,107 flight cycles: much too early for a predicted life. The investigating team came to a conclusion that the cause of the accident was the fatigue fracture of a fuse pin—part of a pylon mounting. While the failed fuse pin could not be recovered, another pin of the same installation was found. The second fuse pin had an internal fatigue crack 4 mm deep, developed from multiple sites in poor quality machining grooves.

The engines of Boeing 747 aircraft are supported by pylons mounted to the wing structure. In the design, each pylon mounting (Figure 8.1) had four fuse pins.[2,3] In the past, cracked fuse pins 2, 3 and 4 have been reported in other Boeing 747 airplanes as well. A fuse pin assembly is shown in Figure 8.2. As stated in the report (Reference 1), the parts during assembly are treated with a corrosion preventive compound. Nevertheless, past inspections of fractured pins revealed existence of corrosion.

8.1.2 Problem analysis

Load acting upon the fuse

Let us consider the fluctuating load acting upon the fuse assembly between maximum and minimum values as per frequency of flight cycles: the maximum load occurs at take-off, while the minimum load takes place when the airplane is at rest. The

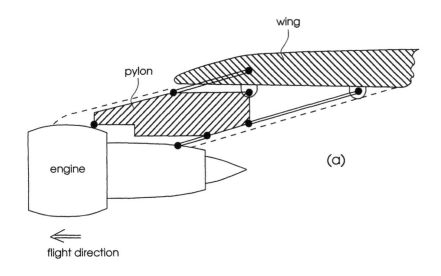

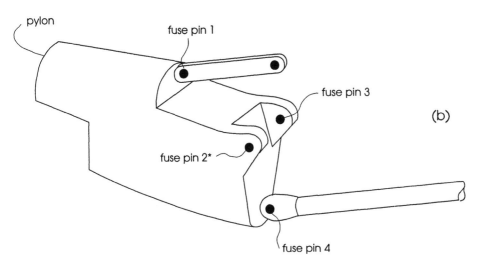

Figure 8.1 Pylon to wing attachement: (a) general scheme; (b) fuse pins. Asterisk (*) indicates a fuse pin that failed.

maximum load comprises two components: the take-off thrust of the engine and the gyroscopic effect as the plane executes a maximum climb. The minimum load pertains to the weight of pylon and engine. We neglect the weight and assume the minimum load to be zero.

Take-off thrust The pylon-wing assembly is mounted on four supports, see Figure 8.1(b). In order to compute the loads, data concerning pylon stiffness are necessary. To replace unavailable data, let us make the following assumption. Based on the fact noted above that fuses 2, 3 and 4 were reported to fail in the past, and since fuse 1 is not known ever to fail, we assume this fuse to be redundant.

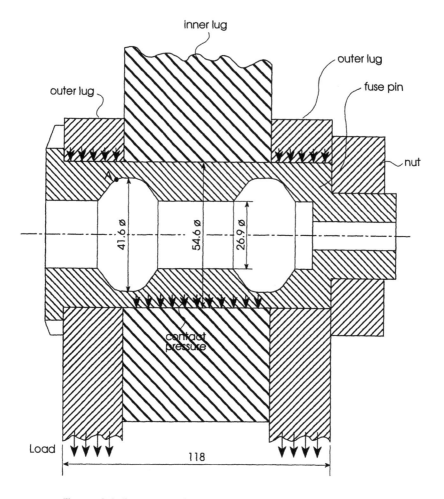

Figure 8.2 Fuse assembly.

The computed reaction forces are indicated in Figure 8.3. Taking into consideration three supports and a take-off thrust of 222.5 kN as listed for Pratt and Whitney engine JT9D-7J,[4] the force acting on fuse 2 is

$$F = \frac{\sqrt{319.1^2 + 218.2^2}}{2} = 193.2 \text{ kN}$$

Gyroscopic effect   The airplane pitching motion at the maximum climb, due to the diverging thrust direction, imposes an additional load upon the pylon-wing mounting, recognized as the gyroscopic effect. It is known from experience that this additional loading amounts to about 40 percent of the thrust force. Thus, using a multiplication factor of 1.40, a total load at fuse 2 assembly becomes

$$F = 1.40 \times 193.2 \text{ kN} = 270.6 \text{ kN}$$

which is equal to 60,800 lbs.

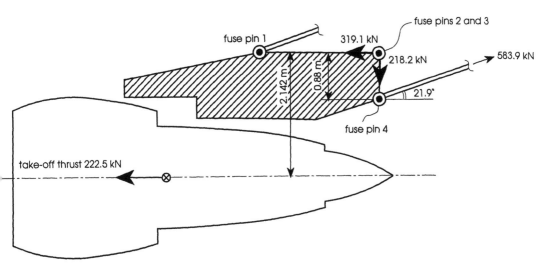

Figure 8.3 Reaction forces in fuse pins.

FE analysis

Figure 8.4 presents an FE model of the fuse assembly. Because of symmetry only one quarter of the fuse assembly is analyzed. The upper surface of the inner lug is fixed. The load is applied at the bottom of the outer lug, as a constant negative pressure, and is transmitted through the pin to the outer lug by surface contact, as was shown in Figure 8.2. In the FE model the surface contact is simulated by gap elements. Figure 8.5 shows the fuse pin itself.

Figure 8.4 FE model of fuse assembly.

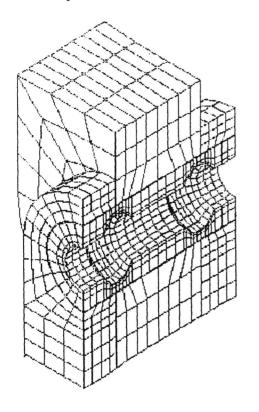

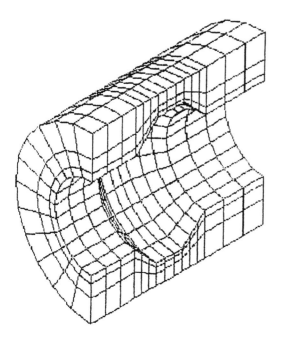

Figure 8.5 FE model of fuse pin.

The FE analysis takes into consideration the elastic-plastic behavior of the material. The original material which was steel SAE 4330M, BHN about 240, was unavailable to us. We chose, instead, the properties of steel SAE 4340, BHN = 243, which were as follows (see Appendix):

Ultimate tensile strength	$S_u = 827$ MPa
Modulus of elasticity	$E = 193,000$ MPa
Yield point (cyclic)	$S_y = 455$ MPa
Cyclic strain hardening exponent	$n' = 0.18$
Fatigue strength coefficient	$\sigma'_f = 1200$ MPa
Fatigue strength exponent	$b = -0.095$
Fatigue ductility coefficient	$\varepsilon'_f = 0.45$
Fatigue ductility exponent	$c = -0.54$

Figure 8.6 shows the stress-strain correlation derived from the above data.

Solution

Because of the elastic-plastic behavior of the material and due to gap elements, the solution proceeds in small load increments using the Newton-Raphson iterative method. The solution is obtained using ANSYS and MSC/NASTRAN programs (see Appendix). The results of the analysis are presented in Figures 8.7 trough 8.9. Figures 8.7 and 8.8 present the distribution of the first principal stress S_1 in the fuse pin. The critical point is at the upper corner at the internal wall surface, point MX, where the stress reaches the value 394.4 MPa. Figure 8.9 shows the stress history of three principal stresses, S_1, S_2 and S_3, in this location. The results show that the stresses remain in the elastic domain.

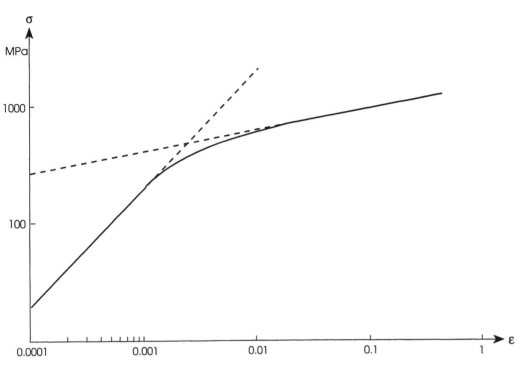

Figure 8.6 Stress-strain correlation of steel SAE 4340, BHN=243.

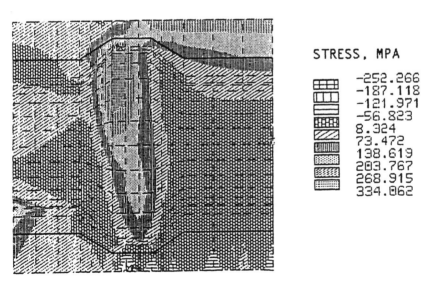

STRESS, MPA

⊞	−252.266
⊞	−187.118
⊞	−121.971
▦	−56.823
▨	8.324
▦	73.472
▦	138.619
▦	203.767
▦	268.915
▦	334.062

Figure 8.7 Stress distribution (principal stresses S_1) in fuse pin—longitudianal view. At maximum point $S_1 = 394.4$ MPa.

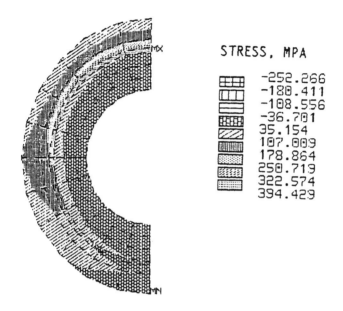

Figure 8.8 Stress distribution (principal stress in S_1) in fuse pin—cross sectional view.

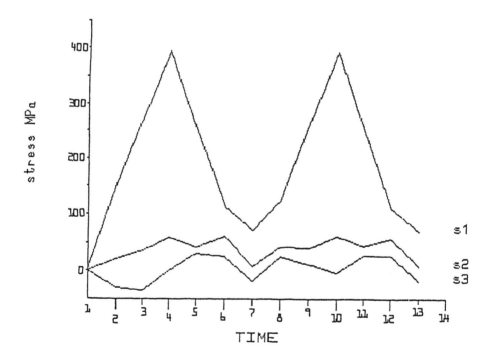

Figure 8.9 Stress history at the critical location in the fuse pin—principal stresses versus time.

Based on the above fact, the life expectancy may be computed using the stress method. According to Figure 8.9, principal stresses S_2 and S_3 are approaching zero. The mean stress, based on S_1, equals

$$S_m = \frac{S_{max} + S_{min}}{2} = \frac{394 + 70}{2} = 232 \text{ MPa}$$

The stress amplitude is

$$S_a = \frac{S_{max} - S_{min}}{2} = \frac{394 - 70}{2} = 162 \text{ MPa}$$

The fatigue limit of the material, per Equation (3.14) and Figure 3.15, equals

$$S_f = 1.0 \times 0.85 \times 0.8 \times \frac{827}{2} = 281 \text{ MPa}$$

Using Equation (3.63) which represented a constant life diagram, see Chapter 3, we can compute the equivalent stress amplitude

$$S_N = 827 \frac{162}{827 - 232} = 225 \text{ MPa}$$

From the fatigue life correlation, Equation (3.60), one gets the life expectancy

$$N = 1000 \left(\frac{0.9 S_u}{S} \right)^{\frac{3}{\log \frac{0.9 S_u}{S_f}}} = 1000 \left(\frac{0.9 \times 827}{225} \right)^{\frac{3}{\log \frac{0.9 \times 827}{281}}} = 4.8 \cdot 10^6 \text{ cycles} \tag{8.1}$$

which indicates practically an infinite life.

8.1.3 Crack propagation

While the stress method indicated no danger of a failure, the crack propagation theory using the same data as above, produces different results. Reference 1 mentions a poor quality surface inside a fuse pin, see above. If we assume a similar surface in the pin that failed, this fact in combination with high tensional stresses could explain the crack propagation leading to fatigue failure. There are two kinds of surface defects that could be considered—a corrosion pit or a machining flaw.

Corrosion

Parts like the fuse assembly are open to unfavorable environment, stemming from excessive temperature and humidity changes, aggravated by leakages of exhaust fumes. The most damaging results of this inevitable phenomenon is corrosion. The only guard against it is to manufacture the parts from corrosion resistant materials. The fuse pin is made of steel SAE 4330M, which does not have anti-corrosive properties and, in spite of preventive measures, the presence of corrosion was noted. See Reference 1. The duration of pin's service, in this case, is sufficient not only for corrosion to set in, but also to cause corrosion pits, reaching depths of 1 mm and more.

Let us approach our investigation by determining in an indirect way the size of a corrosion pit that could have caused the failure when it occurred. After having performed several iteration, by trial and error, we determined a depth of corrosion pit that could cause the fracture. The size of the corrosion pit equals

$$a_i = 0.475 \text{ mm} = 0.000475 \text{ m}$$

For this size, using the tension stress of 394.4 MPa (Figure 8.9), the stress intensity factor equals

$$K_I = 1.12 \times 394.4 \sqrt{\pi \times 0.000475} = 17.06 \text{ MPa}\sqrt{m}$$

which is sufficient for crack propagation. If we assume the crack length at the instance of failure to be equal to the wall thickness of the fuse pin (Figure 8.2), then

$$a_f = 6.5 \text{ mm} = 0.0065 \text{ m}$$

It follows that the fatigue life, as per Equation (5.92), equals

$$N_f = \frac{1}{(n-2)C\left(\alpha S_{\max}\sqrt{\pi}\right)^n} \left[\frac{1}{a_i^{\frac{n-2}{2}}} - \frac{1}{a_f^{\frac{n-2}{2}}} \right] \tag{8.2}$$

$$= \frac{1}{6.9 \times 10^{-12}\left(1.12 \times 394.4\sqrt{\pi}\right)^3} \left[\frac{1}{0.000475^{0.5}} - \frac{1}{0.0065^{0.5}} \right] = 10,107 \text{ cycles}$$

which corresponds to the fatigue life of the failed fuse pin under investigation. Arriving at the same number also confirms that the size of corrosion pit of 0.475 mm is indeed sufficient to cause the fracture. Note that stresses in the vicinity of crack may grow in consequence to a changing geometry, making it possible for a smaller pit to cause the failure.

Machining defect

Unrelated to corrosion damage discussed above, a different source for crack propagation is inherent in the surface due to machining. Machining processing creates surface irregularities and imposes residual stresses at the surface—two major factors in surface integrity affecting the fatigue life, see Chapter 6. Surface defects in combination with high residual tension stresses and cyclic loading can initiate crack propagation. Following is the investigation of surface integrity with the view of establishing it as the cause of the occurred failure.

The report of the investigating team (Reference 1) indicates a poor quality of machining. For general geometry in rough processing the anticipated roughness is $R_a = 10$ to $20\ \mu m$. Considering the difficulties of machining an enclosed interior of the fuse pin, the roughness of it may be worse. Undetected defects in the surface caused by chipping of the cutting edge of the tool, in the form of scratches, grooves and deep cuts, may also be present. In addition to roughness, a machining operation, such as turning, produces residual tension stresses in the surface layer throughout a thickness of up to $200\ \mu m$, reaching to about 750 MPa. See Section 6.3.4.

Let us base the investigation of pin's life expectancy on the following data: average surface roughness of $R_a = 15\ \mu m$ and residual tension stresses within a range of 700 MPa at the surface to zero at a depth of $200\ \mu m$ (see Figure 6.28). The ratio of peak-to-valley height to the average-roughness height is

$$\frac{R_{max}}{R_a} = 4 \text{ to } 5$$

see Section 6.1.1. We may assume the starting length of the crack to equal

$$a_i = R_{max} = 5R_a = 75\ \mu m$$

The final length of the crack, at the failure, equals the wall thickness of the pin, that means,

$$a_f = 6.5 \text{ mm}$$

According to the FE solution, above, the maximum principal stress at the critical location is $\sigma_1 = 394.4$ MPa. The total stresses in the surface layer are obtained by superposition of the residual tension stresses and the FE results. See Figure 8.10. (The stresses in the figure were obtained by combining elastic-plastic strains.) To determine the number of loading cycles leading to the failure, we introduce an approximate step-like stress distribution, see Figure 8.10(a). Accordingly, the number of cycles equals

$$N_f = \sum \Delta N_i \qquad [8.3]$$

with

$$\Delta N_i = \frac{1}{6.9 \cdot 10^{-12} \left(1.12\sigma_i \sqrt{\pi}\right)^3} \left[\frac{1}{a_i^{0.5}} - \frac{1}{a_{i+1}^{0.5}} \right] \qquad [8.4]$$

and where

$$\begin{aligned}
&i = 1, &&\sigma_1 = 666 \text{ MPa}, &&a_1 = 75\ \mu m, &&a_2 = 116\ \mu m; \\
&i = 2, &&\sigma_2 = 557 \text{ MPa}, &&a_2 = 116\ \mu m, &&a_3 = 158\ \mu m; \\
&i = 3, &&\sigma_3 = 448 \text{ MPa}, &&a_3 = 158\ \mu m, &&a_4 = 200\ \mu m; \\
&i = 4, &&\sigma_4 = 394.4 \text{ MPa}, &&a_4 = 200\ \mu m, &&a_5 = 6.5 \text{ mm}.
\end{aligned}$$

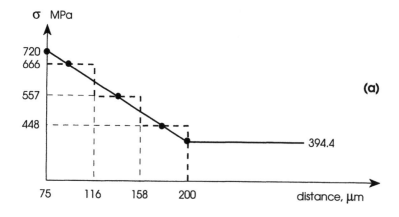

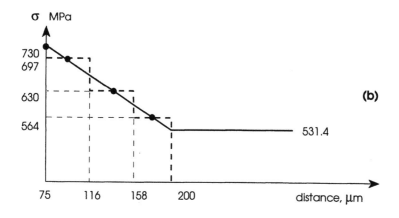

Figure 8.10 Stress distribution at the internal surface of the
fuse pin: (a) normal load; (b) load increased by 24 percent.

Based on the above, the fatigue life equals

$$N_f = 1,419 + 1,425 + 1,822 + 17,607 = 22,273 \text{ cycles}$$

The above result shows that, at the given load level, this kind of defect could not have
caused the fracture at the time it occurred—much sooner than 22,273 cycles.

Increased load   Since the fracture in fact did occur, we shall proceed with the investigation, questioning the magnitude of the load on the fuse assembly that was used in
our computation above. To take into consideration a possibility that the load could have
been greater, let us, thus, by an indirect method (bearing in mind the number of flight
cycles when the failure indeed occurred) arrive at a corresponding value of the load.
After having performed a few iterations, we found that with a load increase of 24

percent the computed number of cycles comes very close to the number of flights when the failure occurred.

Let us follow through the derivation of the number of cycles to failure based on the new assumption. The load upon the fuse assembly equals

$$F = 270.6 \times 1.24 = 335 \text{ kN}$$

A nonuniform load distribution at the outer lug, slanted toward the inner edge is also assumed (see Figure 8.2). Utilizing the same computer procedure as above, a maximum principal stress at the critical location is obtained which equals 531.4 MPa. The total stresses are obtained by superposition of the latter with residual tension stresses, as shown in Figure 8.10(b). The number of cycles to failure can be computed by the equation

$$N_f = \sum \frac{1}{6.9 \cdot 10^{-12} \left(1.12 \sigma_i \sqrt{\pi}\right)^3} \left[\frac{1}{a_i^{0.5}} - \frac{1}{a_{i+1}^{0.5}} \right] \qquad [8.5]$$

with

$i = 1,$	$\sigma_1 = 697$ MPa,	$a_1 = 75 \,\mu m,$	$a_2 = 116 \,\mu m;$
$i = 2,$	$\sigma_2 = 630$ MPa,	$a_2 = 116 \,\mu m,$	$a_3 = 158 \,\mu m;$
$i = 3,$	$\sigma_3 = 564$ MPa,	$a_3 = 158 \,\mu m,$	$a_4 = 200 \,\mu m;$
$i = 4,$	$\sigma_4 = 531.4$ MPa,	$a_4 = 200 \,\mu m,$	$a_5 = 6.5$ mm.

Based on the above, now the fatigue life equals

$$N_f = 1,238 + 985 + 913 + 7,198 = 10,334 \text{ cycles}$$

The number of cycles thus obtained agrees with the number of actual flights that preceded the accident. A machining-damaged surface, coupled by an increased load can therefore be the determining factor in causing the failure.

8.1.4 Conclusion

Fatigue life computed, using the stress method, on the basis of design data of the fuse pin indicates no danger of fracture after 10,107 flights. Satisfied with the result thus obtained, the investigation turned to another factor most affecting fatigue life—the surface integrity of the part. Fatigue failure begins in most cases at the surface which bears the greatest load and is exposed to environmental effects. Using the crack propagation theory a hypothesis was checked to determine whether a defect, caused by either corrosion or by machining, could have been responsible for the failure. As it appears from our analysis, a local surface defect inside the fuse pin could indeed have been the source of a fatigue crack which propagated and caused the critical fracture.

8.2
CRACK PROPAGATION IN A BREECH BLOCK

The purpose of this investigation is to explore the effects of crack propagation on the fatigue life of a breech block in an artillery gun. The analysis shows that a backfire action produces local compression causing plastic deformation. As a result residual tensional stresses develop with a subsequent crack propagation which as the analysis shows, however, will cease when the crack reaches an underformed domain. A conclusion can be reached that, because the process is interrupted, the effect on life expectancy is not as severe as it would be if the propagation were to continue.

8.2.1 The problem

The breech block is a part of the breech mechanism in an artillery gun. The gun is of a kind known as Rheinmetal 120mm gun. Figure 8.11 presents schematically the action of the breech block. With each firing, the block is subject to a high pressure backfire (up to 6700 bar) acting on an exposed area (180mm diameter), see Figure 8.12. The geometry of the breech block is presented in Figure 8.13. The breech block has side grooves to accommodate a mechanism for vertical drive. It is made of steel forging SAE 4340, heat treated to $R_c = 38$.

In the breech blocks under consideration, cracks appeared in corners of the side grooves after about 5000 firings. Additional information given and taken into consideration is the fact that, under backfire, the groove region is subject to compressive stresses. This was observed by painting the surfaces with a photo-elastic substance prior to firing. The objective of the investigation is to evaluate the detrimental effect of the discovered cracks on fatigue life being aware of occurrence of compressive stresses.

8.2.2 FE analysis

The following analysis comprises two parts: (a) a determination of residual stresses; and (b) investigation of crack generation and propagation.

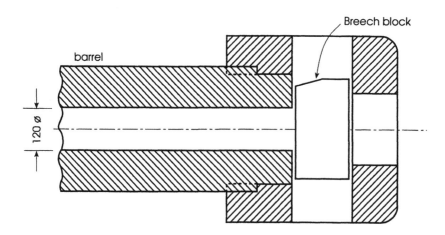

Figure 8.11 Breech mechanism in an artilery gun.

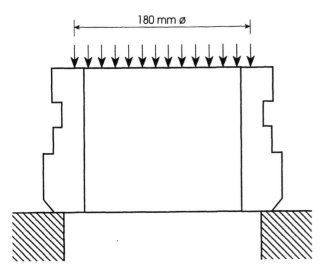

Figure 8.12 Loading of breech block.

Consider FE modeling in a situation where a standard submodeling is not applicable because of the existing plastic deformation. A multistep solution overcomes two constraints: first, a possible computer limitation in trying to achieve the necessary accuracy; and second, the problem of elastic-plastic condition. A coarse FE model is followed by a restricted refined model with the same boundary conditions. The first solution assumes the metal to be purely elastic and the final step takes into consideration the elastic-plastic condition using ANSYS and MSC/NASTRAN programs (see Appendix).

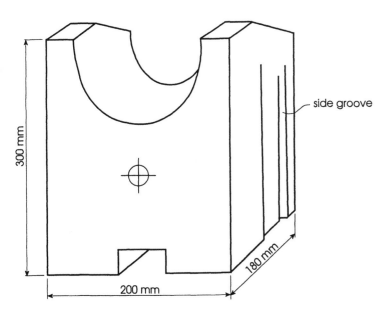

Figure 8.13 Geometry of breech block.

The metal properties used in the analysis are as follows (see Appendix):

Brinell hardness	BHN	= 350
Ultimate tensile strength	S_u	= 1241 MPa
Modulus of elasticity	E	= 193,000 MPa
Yield point (cyclic)	S_y	= 758 MPa
Cyclic strain hardening exponent	n'	= 0.14
Fatigue strength coefficient	σ'_f	= 1655 MPa
Fatigue strength exponent	b	= −0.076
Fatigue ductility coefficient	ε'_f	= 0.73
Fatigue ductility exponent	c	= −0.62

Multistep solution

Elastic condition   Figure 8.14 shows a three-dimensional coarse model of the breech block used in first computation, ignoring some geometrical details. Since the FE model and the loading are symmetrical, we use the half-model shown in Figure 8.15 as the effective model. The model is made of hexahedron elements and is supported at its edges (Figure 8.12). The supports are represented by a set of gap elements. The solution proceeds in steps, starting with the three-dimensional coarse model and ending with a two-dimensional refined model. The first results, in the form of stress intensities, are

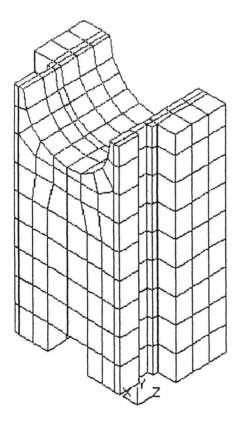

Figure 8.14 Three-dimensional FE model of a breech block.

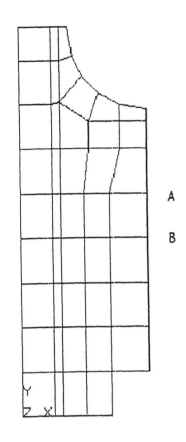

Figure 8.15 FE half model.

shown in Figure 8.16: the stress intensity distribution in two horizontal cross sections, AA and BB, assuming a purely elastic material. The results confirm that the maximum stresses occur, as expected, at the side groove. Based on the fact that the stress distributions are substantially identical, see Figures 8.16(a) and 8.16(b), a two-dimensional coarse model, defined as a flat plate between sections AA and BB, is now used. Figure 8.17 shows the stress intensity distribution in the new model.

Employing the above results, a more precise two-dimensional model is developed where in the vicinity of the grooves a finer meshing is applied. See Figure 8.18. Figure 8.19 presents a more exact solution in enlarged form, still true for elastic condition. Due to assumed purely elastic material and the disregard for a yield point, the obviously wrong stress intensities in groove corners (above 3000 MPa) are a clear indication of plastic deformation. Therefore a more realistic solution is in order taking this into consideration.

Plastic deformation   For the final solution, the same refined FE model is used. The material properties of SAE 4340, as listed above, are reflected in the stress-strain curve plotted in Figure 8.20. The computation takes into consideration a true load cycling, that is, a rising loadingyyy caused by backfire and the unloading after the pressure from backfire subsides. (A listing of computer commands is given in Appendix.) The results of the computation are presented in Figures 8.21 to 8.24. Note that at the end of a loading cycle the stress intensity at the groove reaches plastic condition (Figure 8.21) which upon unloading causes residual tension stresses (Figure 8.22). The stress and strain histories are presented in Figures 8.23 and 8.24, respectively.

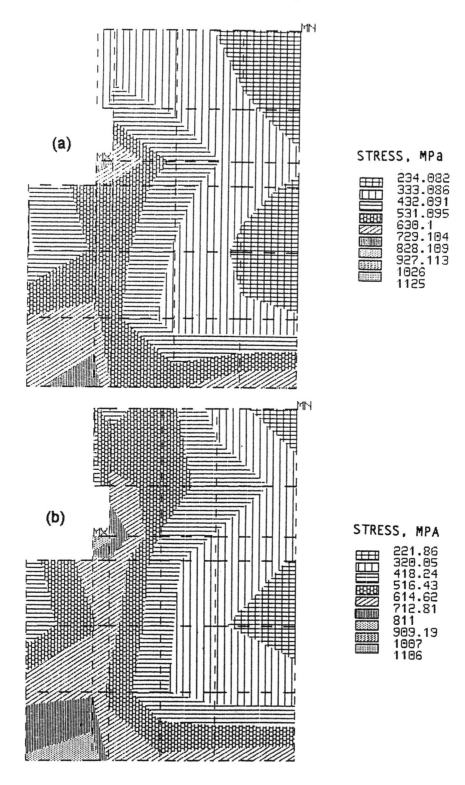

Figure 8.16 Stress intensity distribution in a three-dimensional FE model:
(a) cross section AA; (b) cross section BB.

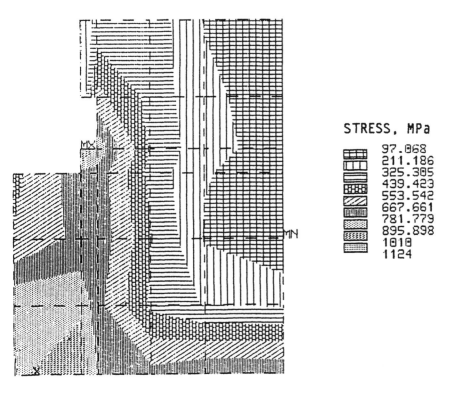

STRESS, MPa
```
97.068
211.186
325.305
439.423
553.542
667.661
781.779
895.898
1010
1124
```

Figure 8.17 Stress intensity distribution in a two-dimensional FE model as per elastic solution.

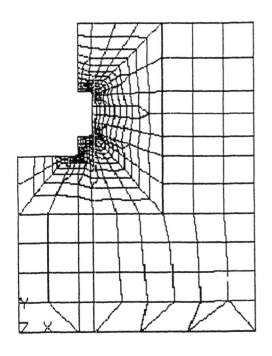

Figure 8.18 Two dimensional FE model with a refined mesh.

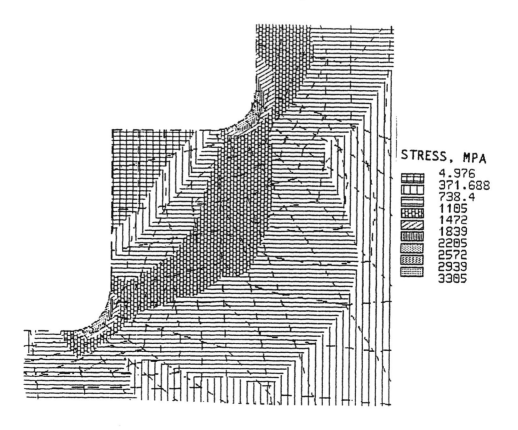

Figure 8.19 Stress intensity distribution in a refined FE model as per elastic solution.

Fatigue   The life expectancy is computed using the strain method. The equivalent stresses and strains at the groove are determined using the criteria of Sines and von Mises. The equivalent mean stress, as per Sines, Equation (4.72), equals

$$\sigma_{m,1} + \sigma_{m,2} + \sigma_{m,3} = 327 - 150 - 585 = -408 \text{ MPa}$$

where the stress values were taken from Figure 8.23. The equivalent alternating strain amplitude, as per von Mises, Equation (4.73), equals

$$\frac{\Delta\varepsilon}{2} = \frac{\sqrt{2}}{3}\left[2\left(\frac{\Delta\varepsilon_1}{2}\right)^2 + 2\left(\frac{\Delta\varepsilon_2}{2}\right)^2 - 2\frac{\Delta\varepsilon_1}{2}\frac{\Delta\varepsilon_2}{2}\right]^{\frac{1}{2}} \qquad [8.6]$$
$$= 0.667(0.00762^2 + 0.011152^2 - 0.0076 \cdot 0.011152)^{0.5} = 0.00657$$

with the strain values taken from Figure 8.24. To derive fatigue life, we use Morrow's procedure, Equation (4.53). The corresponding diagram of the strain amplitude versus fatigue life is shown in Figure 8.25. Accordingly, the fatigue life of the breech block equals

$$N_f = 9000 \text{ cycles}$$

which is the life up to the presence of a visible crack, see Chapter 1.

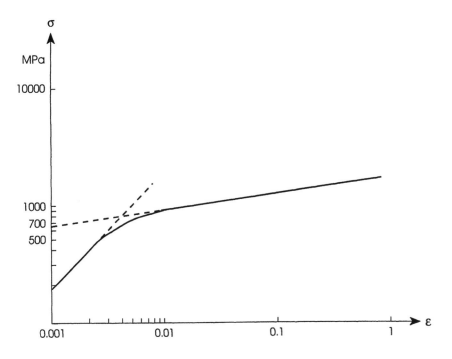

Figure 8.20 Stress-strain correlation of steel SAE 4340, BHN=350.

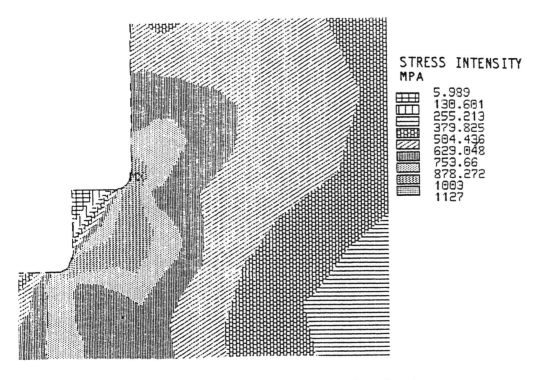

STRESS INTENSITY
MPA

	5.989
	130.601
	255.213
	379.825
	504.436
	629.048
	753.66
	878.272
	1003
	1127

Figure 8.21 Distribution of compression stresses at the end of loading stage.

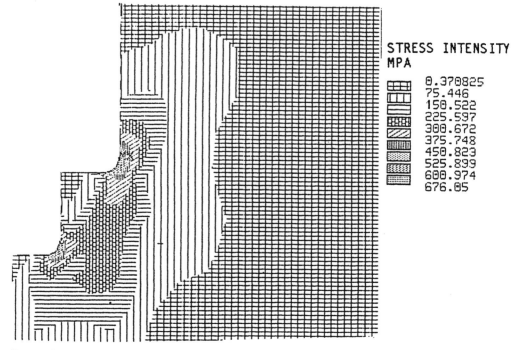

Figure 8.22 Distribution of residual stresses upon unloading.

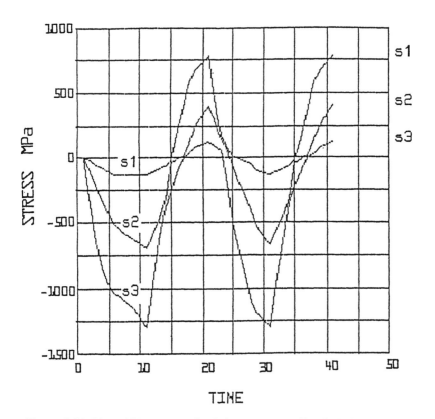

Figure 8.23 Stress history (principal stresses versus time) during loading and unloading.

(x10**-2)

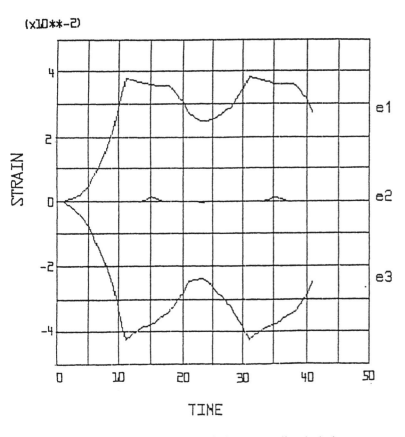

Figure 8.24 Strain history (principal strains versus time) during loading and unloading.

Crack propagation

It follows from the above computation that a crack will start in a corner of the groove after about 9000 loading cycles. This range does not take into consideration a possible existence of surface irregularities which, as we know, may induce a crack propagation at an earlier stage.

Consider the effect of surface irregularities produced by machining. A standard roughness produced by milling operation is in the range from 0.2 to 25 μm. See Chapter 6. The critical tension stress in the corner equals 750 MPa (Figure 8.23). If one assumes an irregularity of 25 μm, the corresponding stress intensity factor at this point equals

$$K_I = 1.12 \times 750 \sqrt{\pi \times 0.000025} \cong 7.5 \text{ MPa}\sqrt{m}$$

This value is sufficiently large to allow a crack propagation much earlier, theoretically from the first cycle.

Within the above assumptions that confirm the existence of crack propagation, we proceed to compute the stress intensity factor of the crack as it propagates beyond the zone of residual stresses. Figure 8.26 shows a FE model with an inserted crack, propagating from the location of the highest residual stresses and beyond, reaching a depth

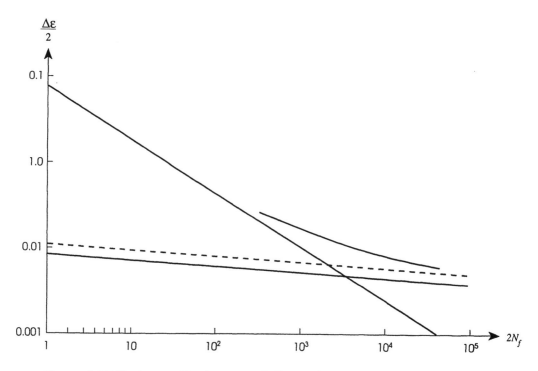

Figure 8.25 Strain amplitude versus fatigue life for steel SAE 4340, BHN=350.

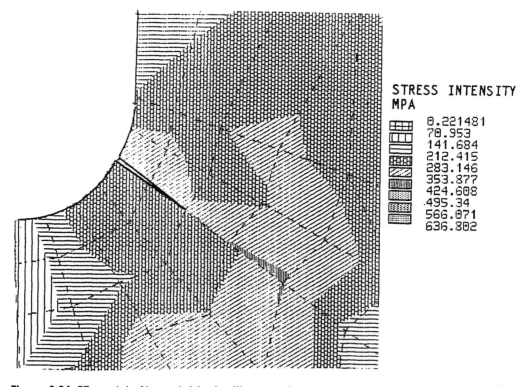

Figure 8.26 FE model of breech block with a crack.

of 3.3 mm. The resultant value of stress intensity factor at this depth, as generated by the computer, equals

$$K_I = 0.5 \text{ MPa}\sqrt{m}$$

The above is much below the threshold value of steel SAE 4340 (see Appendix) which indicates that the crack will not propagate beyond this depth.

8.2.3 Conclusion

The case deals with a cyclic process which fluctuates between the two conditions: tension that induces crack propagation and compression which in general diminishes it. The case under investigation did show an aborted crack propagation, however the presence of tensional stresses causes a controversy. When a local compression in the breech block (during the backfire stage) is high enough to pass the yield point, it causes plastic deformation. This fact is the cause of tensional stresses. Cracks develop after the residual tension stresses reach a critical magnitude and the crack propagation ceases when the tension stresses diminish.

8.3
THERMAL FATIGUE OF A COKE DRUM

The purpose of this investigation is to analyze the fatigue life of a coke drum. The coke drum is used in a delayed coking process to convert hot oils into coke. In the process, the wall of the coke drum is subject to a sequence of extreme heating followed by cooling shocks, causing severe thermal fatigue of the material with consequent cracking of the walls. The conclusion of the following analysis is that the problem may be attributed to the severe heat and mass transfer that takes place at the inner side of the wall. The given condition is outside of design solution.

8.3.1 The problem

As a part of oil refining process, the majority of oil refineries use a delayed coking process to convert heavy oils into coke using coke drums. See Figure 8.27. The refinery process involves filling a coke drum with hot oil, removing the vaporized products and forming coke from the remaining liquid products. The thermal cycle is illustrated in Figure 8.28. After vapor preheating of the coke drum, hot oil enters from the bottom at a temperature of about 900°F. Following a relatively long period during which coke builds up, water is injected to quench the coke. The water quench is followed by removal of coke using high pressure water jets. The total operation takes about 20 to 40 hours.

The coke drum under consideration is made from Cr-Mo steel plate, SA 387 B (grade 11 class 1), by welding. The drum is covered by a layer of mineral wool, 6 in. thick, for thermal insulation. The vapor preheat and water quench are sources of strong thermal shocks having an adverse effect on the drum material. Due to the thermal stresses, the coke drum is subject to severe bulging and cracking. As a rule, after about

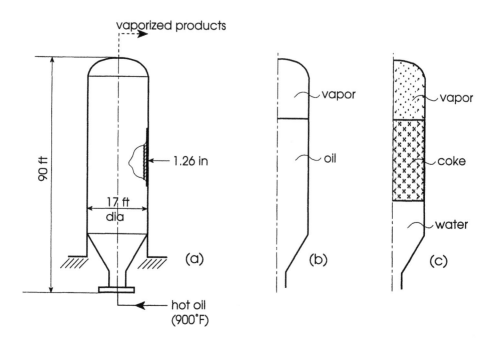

Figure 8.27 Delayed coking process; (a) coke drum configuration
(b) oil fill; (c) water quench.

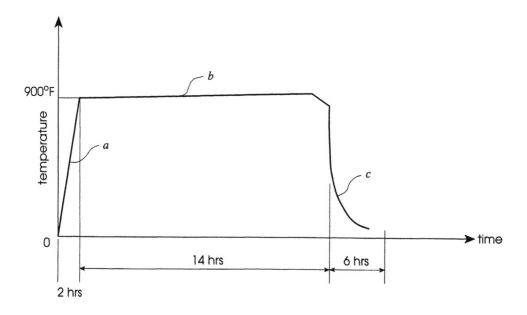

Figure 8.28 Thermal cycle of the delayed coking process; (a) vapor preheat;
(b) oil fill; (c) water quench.

ten years of service, circumferential cracks begin to appear in the cylindrical wall, starting from the inside. Typically the cracks occur at lower level circumferential welds.

Thermal stresses

The vapor preheat and water quench cause sudden temperature changes in the wall, inducing steep temperature gradients and high thermal stresses. The thermal stresses are derived from the modified Hooke's relation

$$\sigma_x = \frac{1}{E}\left[\sigma_x - \nu\left(\sigma_y + \sigma_z\right)\right] + \alpha\Delta T \tag{8.7}$$

$$\sigma_y = \frac{1}{E}\left[\sigma_y - \nu\left(\sigma_z + \sigma_x\right)\right] + \alpha\Delta T$$

$$\sigma_z = \frac{1}{E}\left[\sigma_z - \nu\left(\sigma_x + \sigma_y\right)\right] + \alpha\Delta T$$

where $\alpha\Delta T$ denotes the thermal expansion of the material under free conditions when the temperature is raised by ΔT. Computation of the wall temperature involves the solution of a transient heat transfer problem expressed by the partial differential equation

$$pc\frac{\partial T}{\partial t} = k\left(\frac{\partial^2 T}{\partial \gamma^2} + \frac{1}{\gamma}\frac{\partial T}{\partial \gamma}\right) \tag{8.8}$$

with boundary conditions at the surface of the wall

$$k\frac{\partial T}{\partial \gamma} = h\left(T - T_a\right) \tag{8.9}$$

and an initial condition $T = T_{init}$, where k denotes the thermal conductivity of the material and h is the coefficient of heat transfer at a surface.[5] The main task at this stage is a determination of h which depends on the conditions at the inner and outer wall surfaces.

Consider the conditions at the inner surface of the wall first. Our analysis is limited to the effects of vapor preheat and water quench only, ignoring the oil fill, the phase where the influence of heat transfer is considerably smaller.

Vapor preheat   The process pertains to preheating the wall to about 900°F using water vapor. It involves heat transfer to the wall by free convection in a turbulent environment inside the drum. The heat transfer coefficient[6] in the prevailing conditions at the wall equals

$$h = 0.103\frac{k}{l}\sqrt[3]{Ra} \tag{8.10}$$

where Ra is the Rayleigh number defined by the equation

$$Ra = \frac{g\beta\Delta T l^3}{\nu\alpha} \tag{8.11}$$

l denotes a characteristic length, ΔT is a characteristic temperature difference, β is the thermal expansion coefficient, v is the viscosity, and α is the thermal diffusivity—all relating to the ambient medium. Figure 8.29 presents the heat transfer coefficient h as a function of wall temperature per Equation (8.11). In the present analysis we use a linear approximation shown in the figure.

Water quench　　The thermal effect of quench water upon the wall is difficult to predict, with its chaotic character, particularly when the contact of quench water with wall surface is hindered by coke. Because of the high temperature we are faced with a phenomenon called "surface boiling." For computational purposes we assume that it is unaffected by the presence of the coke.

Empirical correlation concerning the heat flux during surface boiling is presented graphically in Figure 8.30. (See Reference 6.) In the present temperature range, up to

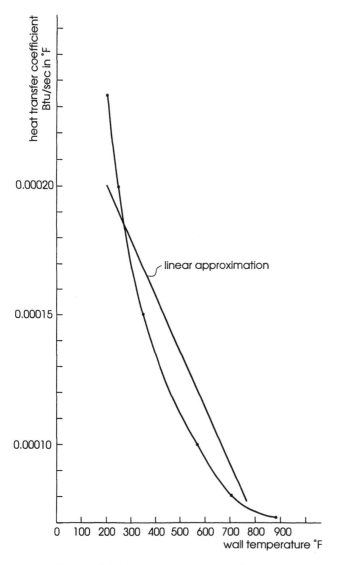

Figure 8.29 Heat transfer coefficient during vapor preheat.

about 900°F, there is a need for extrapolation. Figure 8.31 presents the derived distribution of heat transfer coefficients, including a linear approximation. The approximated heat transfer coefficients are kept below 1 Btu/sec in °F in order to minimize the errors caused by extrapolation.

Outer heat transfer   The heat transfer from the coke drum to the outer air is by free convection along a vertical wall. (See Reference 6.) The heat transfer coefficient is computed according to the equation

$$h_{out} = 0.47 \frac{k}{l} \sqrt[4]{Gr} \qquad\qquad [8.12]$$

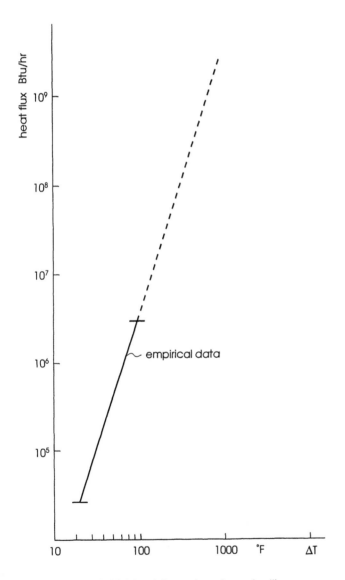

Figure 8.30 Heat flux at surface boiling versus temperature difference at wall.

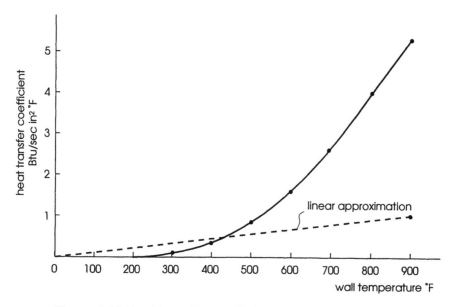

Figure 8.31 Heat transfer coefficient during water quench.

where Gr is the Grashof number defined by the equation

$$Gr = \frac{g\beta\Delta T l^3}{\nu^2} \qquad [8.13]$$

Computing the cooling effect of outer air, one has to consider the heat transfer resistance of the mineral wool covering the wall as well. The combined expression of the outer heat transfer coefficient and the wool resistance is

$$h = \left(\frac{l_{wool}}{k_{wool}} + \frac{1}{h_{out}}\right)^{-1} \qquad [8.14]$$

See Reference 5. This coefficient with the prevailing conditions at the wall equals

$$h = 0.085 \frac{Btu}{sec \cdot in^2 \cdot {}^\circ F}$$

8.3.2 FE analysis

The FE analysis involves the determination of wall temperature and thermal stresses at the same time. The geometrical form and dimensions of the coke drum are shown in Figure 8.27(a). The analysis here is restricted to the cylindrical part of the coke drum. It is also assumed, for simplicity, that the loading is constant throughout the height of the cylinder. It is also assumed that the wall thickness (1.26 in.) is constant. Because of axial symmetry, the analysis is two-dimensional. Figure 8.32 shows the FE model used in the analysis.

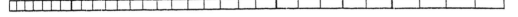

Figure 8.32 Finite element model of the wall of a coke drum.

The numerical analysis takes into consideration the true elastic-plastic behavior of the material. The material properties of Cr-Mo steel plate, SA 387 B,[7] are as follows:

	at room temperature	at 900°F
Ultimate tensile strength	$S_u = 72$ ksi	64 ksi
Modulus of elasticity	$E = 30,000$ ksi	25,000 ksi
Yield point	$S_y = 35.5$ ksi	27.2 ksi
Cyclic strain hardening exponent	$n' = 0.14$	
Fatigue strength coefficient	$\sigma'_f = 122$ ks	114 ksi
Fatigue strength exponent	$b = -0.12$	
Fatigue ductility coefficient	$\varepsilon'_f = 0.20$	
Fatigue ductility exponent	$c = -0.6$	

(Some data, taken from other Cr-Mo steels, were added.) Figure 8.33 shows the stress-strain correlation, derived from the above data.

Solution

The loading on the drum wall originates from several sources: transient heating and cooling, internal pressure, and the weight of the drum. Owing to the transient charac-

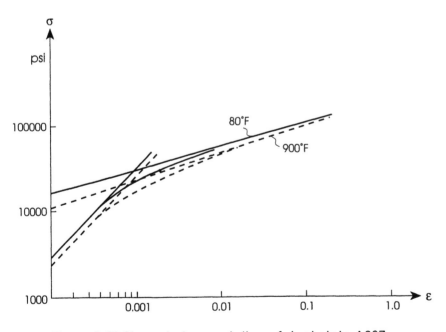

Figure 8.33 Stress-strain correlation of steel plate A387.

ter of the problem, the solution proceeds in time steps, using the Newton-Raphson iterative method at each step because of elastic-plastic behavior of the material.

The stress and strain distribution is achieved using the FE method using ANSYS program (for input see Appendix). The results of computation are presented in Figures 8.34 to 8.37. Figure 8.34 shows the wall temperature, while Figure 8.35 presents the stresses at the internal wall surface, both as functions of time. Figures 8.36 and 8.37 show the stress-strain loops of circumferential and axial stresses and strains at the internal wall surface. (Note: the axial and circumferential stresses are the principal stresses.)

Having obtained the stresses and strains, the life expectancy can be derived using the equivalent stresses and strains as discussed in Section 4.2 dealing with multiaxial analysis. The equivalent mean stress is computed according to Sines criterion, Equation (4.72), using the stress data shown in Figures 8.36 and 8.37,

$$\sigma_{m,1} + \sigma_{m,2} = \sigma_{m,\text{ax}} + \sigma_{m,\theta} = 12,000 + 11,000 = 23,000 \text{ psi}$$

The equivalent alternating strain amplitude, computed according to von Mises criterion, Equation (4.73), equals

$$\frac{\Delta\varepsilon}{2} = \frac{\sqrt{2}}{3}\left[2\left(\frac{\Delta\varepsilon_1}{2}\right)^2 + 2\left(\frac{\Delta\varepsilon_2}{2}\right)^2 - 2\frac{\Delta\varepsilon_1}{2}\frac{\Delta\varepsilon_2}{2}\right]^{\frac{1}{2}} \tag{8.15}$$

$$= 0.667(0.00302^2 + 0.00312^2 - 0.00302 \cdot 0.00312)^{0.5} = 0.002035$$

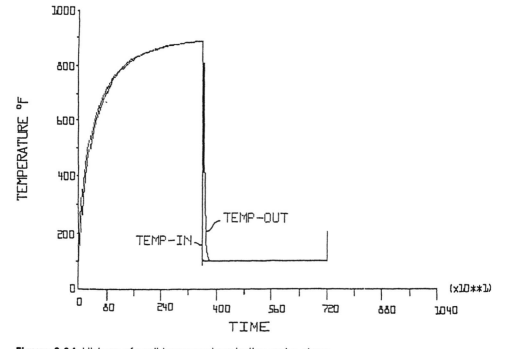

Figure 8.34 History of wall temperature in the coke drum.

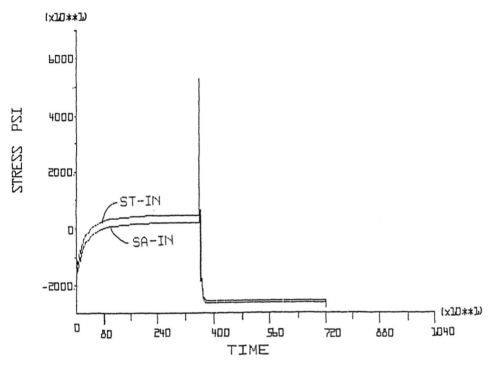

Figure 8.35 History of circumferential and axial stresses at the internal wall surface in the coke drum.

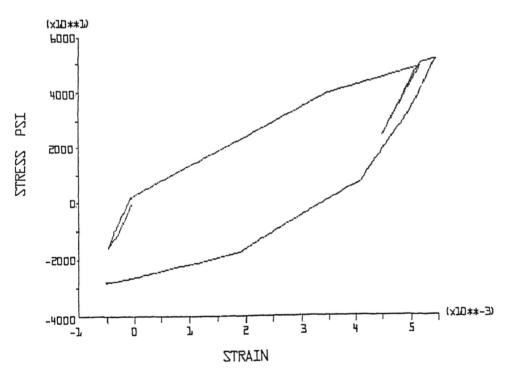

Figure 8.36 Stress-strain loop at the internal surface: axial stresses versus axial strains.

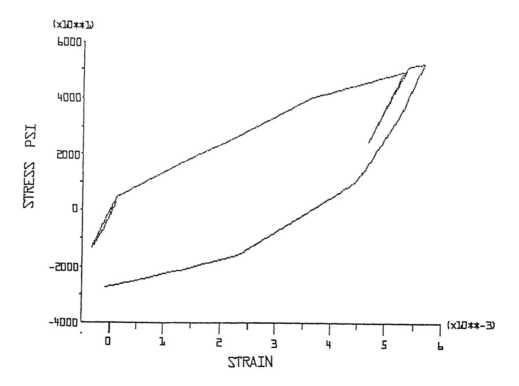

Figure 8.37 Stress-strain loop at the internal surface: circumferential stresses versus circumferential strains.

where the strain values are taken from Figures 8.36 and 8.37. The strain method defines fatigue life up to the occurrence of a visible crack. See Chapter 1. To derive fatigue life here, we use Morrow's procedure, Equation (4.53),

$$\frac{\Delta\varepsilon}{2} = \frac{(\sigma_f' - \sigma_m)}{E}\left(2N_f\right)^b + \varepsilon_f'\left(2N_f\right)^c \qquad [8.16]$$

The corresponding diagram of the strain amplitude versus fatigue life is plotted in Figure 8.38. Accordingly, the fatigue life equals

$$N_f = 4,500 \text{ cycles}$$

which is the life up to the presence of a visible crack, see Chapter 1.

    Residual stresses   In the above analysis the detrimental effects of welding were not taken into consideration. Due to a rapid cooling of a welded seam during the drum's construction, a plastic deformation takes place inducing tensional and compressive residual stresses. Figure 8.39 shows a distribution of the residual stresses along a circumferential weld, in axial direction. It is safe to assume the magnitude of tensional

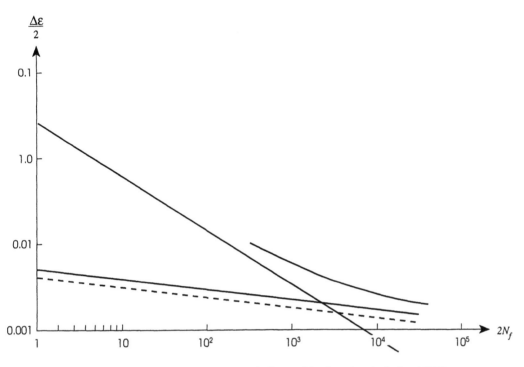

Figure 8.38 Strain amplitude versus fatigue life for steel plate A387.

residual stresses approaches the yield point.[8] In the present case the residual stresses are assumed equal

$$\sigma_{res,ax} = 30{,}000 \text{ psi}$$

Taking now into consideration the residual stresses, we compute anew the stress and strain distribution. Figures 8.40 and 8.41 show the results at the internal wall surface. The equivalent mean stress per Sines criterion in this case equals

$$\sigma_{m,1} + \sigma_{m,2} = \sigma_{m,ax} + \sigma_{m,\theta} = 18{,}500 + 10{,}000 = 28{,}500 \text{ psi}$$

Computing the strain amplitude, let us follow through the more conservative approach above. The equivalent alternating strain amplitude, computed according to von Mises criterion, equals

$$\frac{\Delta\varepsilon}{2} = 0.0020$$

Consequently the life expectancy becomes

$$N_f = 3000 \text{ cycles}$$

If we accept the latter result, it would mean that a safe life of a coke drum is about 3000 cycles.

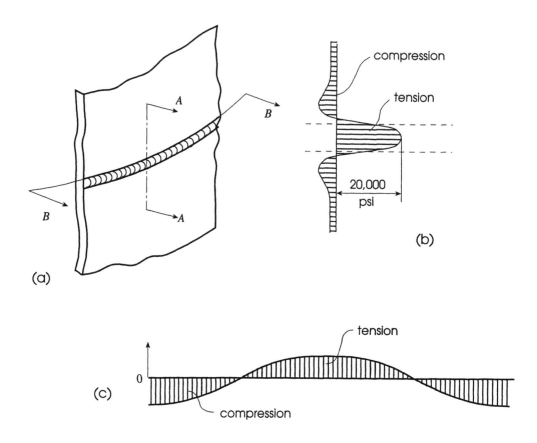

Figure 8.39 Residual stresses in a circumferential weld:
(a) weld configuration; (b) residual stress distribution in axial direction;
(c) residual stress distribution in circumferential direction.

8.3.3 Conclusion

The cause of limited fatigue life of the coke drum is the severe heat and mass transfer process that takes place at the inner side of the wall during the preheating and water quench periods. The prevailing heat transfer coefficients are characteristic of the refining process and are unavoidable facts. Since the cause for failure does not lie in faulty design, the conclusion is reached that a timely replacement or repair is the only solution to the problem. However, a concern for a special weld treatment to reduce residual stresses could improve the life êxpectancy. According to our theoretical investigation, an increase of life expectancy could be as much as 33 percent. (Please note that a change in wall thickness, even though one would expect it to be a factor in the fatigue life analysis, did not prove to be so in subsequent computations.)

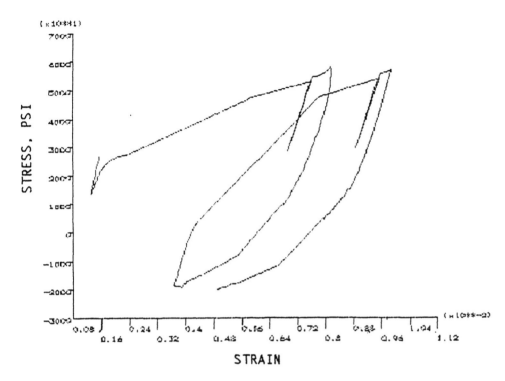

Figure 8.40 Stress-strain loop at the internal surface when residual stresses are present: axial stresses versus axial strains.

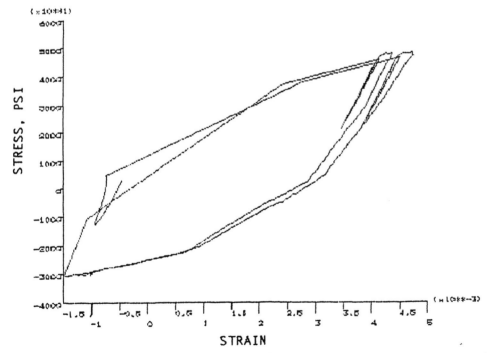

Figure 8.41 Stress-strain loop at the internal surface when residual stresses are present: circumferential stresses versus circumferential strains.

REFERENCES

1. *Aircraft Accident Report 92–11*. 1992. Netherlands Aviation Safety Board.
2. Niu, M.C.Y. 1988. *Airframe Structural Design*. Los Angeles, CA: Technical Book Company.
3. Morrison, J.A. 1968. *747 Engine Installation Features*. SAE 680335. Warrendale, PA: Society of Automotive Engineers.
4. *Jane's All The World Aircraft*. 1980. London: Jane's Publishing Company.
5. Ozisik, M.N., 1993. *Heat Conduction*. New York: Wiley-Interscience.
6. Rohsenow, W.M., J.P. Hartnett and E.N. Ganic, eds., 1985. *Handbook of Heat Transfer Applications*. New York: McGraw-Hill.
7. *Metals Handbook*. 1990. Vol. 1, 10th ed. Metal Park, OH: American Society of Metals.
8. Gurney, T.R. 1979. *Fatigue of Welded Structures*. Cambridge, Great Britain: Cambridge University Press.

APPENDIX

In addition to the reference tables found within the text, the appendix contains five tables in support of theories presented in Chapters 3, 4, and 5, and fatigue cases solved in Chapter 8.

The list of tables follows:

A.1 Stress-Strain Properties of Selected Steels
A.2 Plane-Strain Fracture Toughness of Selected Steels
A.3 Threshold Data for Selected Steels
A.4 Input Data Listing for Fatigue Cases—ANSYS Program
A.5 Input Data Listing for Fatigue Cases—MSC/NASTRAN Program

Note: ANSYS is a registered trademark of SAS IP. MSC/NASTRAN is a trademark of The MacNeal-Schwendler Corporation.

APPENDIX

TABLE A.1
Monotonic and Cyclic Stress-Strain Properties of Selected Steels

SAE	BHN	S_u (MPa)	S_y monotonic (MPa)	S_y cyclic (MPa)	σ'_f (MPa)	ε'_f	n'	b	c	% RA	E (GPa)
1015	80	415	228	241	825	0.95	0.22	−0.11	−0.64	68	207
1020	108	440	262	241	895	0.41	0.18	−0.12	−0.51	62	203
1040	225	620	345	386	1540	0.61	0.18	−0.14	−0.57	60	200
1045	225	725	634	414	1225	1.00	0.18	−0.095	−0.66	65	200
1045	410	1450	1365	827	1860	0.60	0.15	−0.073	−0.70	51	200
1045	390	1345	1276	758	1585	0.45	0.17	−0.074	−0.68	59	207
1045	450	1585	1517	965	1795	0.35	0.15	−0.07	−0.69	55	207
1045	500	1825	1689	1276	2275	0.25	0.12	−0.08	−0.68	51	207
1045	595	2240	1862	1724	2725	0.07	0.13	−0.081	−0.60	41	207
1144	265	930	717	552	1000	0.32	0.15	−0.08	−0.58	33	197
1144	305	1035	1020	565	1585	0.27	0.18	−0.09	−0.53	25	199
4130	258	895	779	565	1275	0.92	0.13	−0.083	−0.63	67	221
4130	365	1425	1358	827	1695	0.89	0.12	−0.081	−0.69	55	200
4140	310	1075	965	621	1825	1.20	0.14	−0.08	−0.59	60	201
4142	310	1060	1048	745	1450	0.22	0.18	−0.10	−0.51	29	200
4142	335	1250	1234	1248	1250	0.06	0.14	−0.08	−0.62	28	199
4142	380	1415	1379	827	1825	0.45	0.17	−0.08	−0.75	48	207
4142	400	1550	1448	896	1895	0.50	0.16	−0.09	−0.75	47	200
4142	450	1750	1586	1069	2000	0.40	0.16	−0.08	−0.73	42	207
4142	475	2035	1896	1103	2070	0.20	0.15	−0.082	−0.77	20	200
4142	450	1930	1862	1069	2105	0.60	0.16	−0.09	−0.76	37	200
4142	475	1930	1724	1344	2170	0.09	0.13	−0.081	−0.61	35	207
4142	560	2240	1689	1724	2655	0.07	0.12	−0.089	−0.76	27	207
4340	243	827	634	455	1200	0.45	0.18	−0.095	−0.54	43	193
4340	409	1470	1372	827	2000	0.48	0.15	−0.091	−0.60	38	200
4340	350	1241	1172	758	1655	0.73	0.14	−0.076	−0.62	57	193
5160	430	1670	1531	1000	1931	0.40	0.15	−0.071	−0.57	42	193
9262	260	925	455	524	1040	0.16	0.15	−0.071	−0.47	14	207
9262	280	1000	786	648	1220	0.41	0.12	−0.073	−0.60	33	193
9262	410	1565	1379	1048	1855	0.38	0.09	−0.057	−0.65	32	200
30304	160	745	255	717	2413	1.02	0.36	−0.15	−0.69	74	186
30304	327	951	745	876	2275	0.89	0.17	−0.12	−0.77	69	172
30310	145	641	221	345	1655	0.60	0.26	−0.15	−0.57	64	193

Source: *SAE Handbook.* 1989. Warrendale, PA: Society of Automotive Engineers.

Appendix

TABLE A.2
Plane-Strain Fracture Toughness of Selected Steels

Steel	Properties	K_{Ic} (MPa $m^{0.5}$)
4340	BHN = 380	110
	BHN = 430	75
	BHN = 550	53
4340	air melted	45
	vacuum arc remelted	60
H13 longitudinal direction	$R_c = 40$	78
	$R_c = 46$	52
	$R_c = 50$	33
	$R_c = 53$	24
HP 9–4–30	$R_c = 44–48$	99–115
	$R_c = 49–53$	66–99
300M air melted	longitudinal direction	49
	transverse direction	59
300M vacuum arc remelted	longitudinal direction	57
	transverse direction	64
Tempered martensite	UTS = 1500MPa	80–120
	UTS = 1750MPa	50–100
	UTS = 2000MPa	30–70
Maraging steels		
18Ni(200)	UTS = 1500MPa	155–200
18Ni(250)	UTS = 1800MPa	120
18Ni(300)	UTS = 2050MPa	80
18Ni(350)	UTS = 2450MPa	35–50
18Ni(cast)	UTS = 1750MPa	105

Source: *Metals Handbook.* 1990. Vol. 1, 10th ed. Metal Park, OH: American Society of Metals.

TABLE A.3
Threshold Data for Selected Steels

Steel	$R = K_{min}/K_{max}$	ΔK_{th} (MPa $m^{0.5}$)
Mild steel	0.13	6.6
	0.35	5.2
	0.49	4.3
	0.64	3.2
	0.75	3.8
Low alloy steel	0.0	6.6
	0.32	5.2
	0.5	4.4
	0.62	3.3
	0.73	2.5
9310	0.3	5.5
	0.9	3.3
A508 Class 2	0.1	6.7
	0.5	5.6
	0.7	3.1
A517-F	0.21	5.8
	0.4	4.5
	0.88	3.3
A533 Grade B Class 1	0.1	7.7
	0.3	5.9
	0.5	4.9
	0.8	3.1
18/8 Austenitic	0.0	6.0
	0.33	5.9
	0.62	4.6
	0.74	4.1

Source: Rolfe, S.T., and J.M. Barsom. 1977. *Fracture and Fatigue Control in Structures.* Englewood Cliffs, NJ: Prentice-Hall.

TABLE A.4
Input Data Listing for Fatigue Cases—ANSYS Program*

FUSE PIN

```
/TITLE,fuse pin
/PREP7
C***
MPTEMP,1,0,500
ET,1,45 $MP,EX,1,193E3
MP,NUXY,1,.3 $,DENS,1,7.8E-6
ET,2,52 $R,2,1.4E6,-1E-6
TB,MKIN,1,2
TBTEMP,,STRAIN
TBDATA,1,0.001,0.002,0.005,0.015,0.45
TBTEMP,0.0
TBDATA,1,193.0,320.0,470.0,630.0,1200.0
TBTEMP,500
TBDATA,1,193.0,320.1,470.1,630.1,1200.1
C***
C*** NODES
N, 1,0.00000,27.300,0.00000
N, 2,0.00000,19.304,19.304
N, 3,0.00000,26.370,7.0658
...
...
...

C*** ELEMENTS
E,1,11,13,3,44,49,57,46
E,11,12,14,13,49,51,61,57
E,12,8,9,14,51,23,25,61
...
...
...
TYPE,2
REAL,2
E,120,949
E,122,950
E,124,951
...
...
...
WSORT,X
WSORT,Y
FINISH
C***
/SOLUTION
C***
C*** BOUNDARY CONDITIONS
ACEL,0,386.4,0
NSEL,S,LOC,X,-0.25,0.25
DSYM,SYMM,X
NSEL,ALL
NSEL,S,LOC,Z,-0.25,0.25
DSYM,SYMM,Z
NSEL,ALL
```

```
NSEL,S,LOC,Y,75.54,75.59
D,ALL,ALL
NSEL,ALL

C***
*CREATE,LOAD
NSEL,S,LOC,Y,-76.22,-76.18
SF,ALL,PRES,ARG1
NSEL,ALL
LSWRITE
*END
C***
*DO,I,0,3
*USE,LOAD,I*(-19.479)
*ENDDO
*DO,I,1,3
*USE,LOAD,-58.437+(I*19.479)
*ENDDO
*DO,I,1,3
*USE,LOAD,I*(-19.479)
*ENDDO
*DO,I,1,3
*USE,LOAD,-58.437+(I*19.479)
*ENDDO
LSSOLVE,1,13
FINISH

***
```

BREECH BLOCK

Model: 3-dim coarse
```
/TITLE,breech block (3-dim elastic)
/PREP7
C***
ET,1,45
MP,EX,1,193E3
MP,NUXY,1,.3
MP,DENS,1,7.8E-6
ET,2,52
R,2,1.4E6,-1E-6
C***
C*** NODES
N, 1,000.0,000.0,000.0
N, 2,030.0,000.0,000.0
N, 3,037.5,000.0,000.0
...
...
...
C*** ELEMENTS
E, 1, 2, 7, 6,101,102,107,106
E, 2, 3, 8, 7,102,103,108,107
E, 3, 4, 9, 8,103,104,109,108
```

```
...
...
...
TYPE,2
REAL,2
E,701,601
E,702,602
E,703,603
...
...
...
WSORT,Y
FINISH
C***
/SOLUTION
C*** BOUNDARY CONDITIONS
NSEL,S,LOC,X,120.0
LOCAL,11,0,120.0
DSYM,SYMM,X,11
NSEL,ALL
CSYS
ACEL,0,0.001
NSEL,S,NODE,,701,757
D,ALL,ALL
NSEL,ALL
NSEL,S,NODE,,1001,1605
D,ALL,ALL
NSEL,ALL
ESEL,S,ELEM,,17,22,5
ESEL,A,ELEM,,13,28,5
ESEL,A,ELEM,,14,29,5
NSLE,S
NSEL,R,LOC,Z,-.1,.1
SF,ALL,PRESS,731.54
NSEL,ALL
ESEL,ALL
SOLVE
FINISH
```

Model: 2-dim elastic coarse

```
/TITLE,breech block (2-dim elastic
  coarse)
/PREP7
C***
ET,1,42,,,2
MP,EX,1,193E3
MP,NUXY,1,.3
ET,2,12
R,21,0,0.14E6,1.0E-6
R,22,0,0.065E6,1.0E-6
C***
C*** NODES
N, 1,000.0,000.0
N, 2,030.0,000.0
N, 3,037.5,000.0
...
...
...

C*** ELEMENTS
E, 1, 2,12,11
E, 2, 3,13,12
E, 3, 4,14,13
...
```

```
...
...
TYPE,2
REAL,21
E,101,1
REAL,22
E,102,2
C***
WSORT,Y
FINISH
C***
/SOLUTION
C*** BOUNDARY CONDITIONS
NSEL,S,NODE,,101,102
D,ALL,ALL
NALL
NSEL,S,LOC,X,119.9,120.1
DSYM,SYMM,X,0
NSEL,ALL
NSEL,S,LOC,Y,150.
NSEL,R,LOC,X,37.5,120.
SF,ALL,PRES,385.5
NALL
SOLVE
FINISH
```

Model: 2-dim elastic precise

```
/TITLE,breech block (2-dim elastic
  precise)
/PREP7
C***
ET,1,42,,,2
MP,EX,1,193E3
MP,NUXY,1,.3
ET,1,42,,,2
ET,2,2,,,2
ET,3,12
R,31,0,0.14E6,1.0E-6
R,32,0,0.065E6,1.0E-6
C***
C*** NODES
N,001,30.000,117.00,0.0
N,002,30.000,118.27,0.0
N,003,30.000,119.74,0.0
...
...
...
C*** ELEMENTS
E,009,039,040,010
E,010,040,041,011
E,011,041,042,012
...
...
...
ETYPE,2
E,156,157,146
E,325,312,324
E,397,396,414
...
...
...
TYPE,3
REAL,31
```

```
E,457,456
REAL,32
E,404,403
C***
WSORT,Y
FINISH
C***
/SOLU
C*** BOUNDARY CONDITIONS
D,457,ALL
D,404,ALL
NSEL,S,LOC,X,119.9,120.1
DSYM,SYMM,X,0
NSEL,ALL
NSEL,S,LOC,Y,150.
NSEL,R,LOC,X,37.5,120.
SF,ALL,PRES,385.5
NALL
C***
SOLVE
FINISH
```

Model: 2-dim plastic precise

```
/TITLE,breech block (2-dim plastic
  precise)
/PREP7
C***
ET,1,42,,,2
ET,2,2,,,2
ET,3,12
MPTEMP,1,0,500
MP,EX,1,193E3,1
MP,NUXY,1,.3,0.0001
TB,MKIN,1,2
TBTEMP,,STRAIN
TBDATA,1,0.002590,0.005924,0.079997,0.
  2,0.73
TBTEMP,0.0
TBDATA,1,500,758,1200,1380,1654
TBTEMP,500
TBDATA,1,500,759,1201,1381,1655
R,31,0,0.14E6,1.0E-6
R,32,0,0.065E6,1.0E-6
C***
C*** NODES
N,001,30.000,117.00,0.0
N,002,30.000,118.27,0.0
N,003,30.000,119.74,0.0
...
...
...
C*** ELEMENTS
E,009,039,040,010
E,010,040,041,011
E,011,041,042,012
...
...
...
ETYPE,2
E,156,157,146
E,325,312,324
E,397,396,414
...
...
```

```
...
TYPE,3
REAL,31
E,457,456
REAL,32
E,404,403
C***
WSORT,Y
FINISH
C***
/SOLUTION
C***
C*** BOUNDARY CONDITIONS
D,457,ALL
D,404,ALL
NSEL,S,LOC,X,119.9,120.1
DSYM,SYMM,X,0
NSEL,ALL
C***
*CREATE,LOAD
NSEL,S,LOC,Y,150.
NSEL,R,LOC,X,37.5,120.
SF,ALL,PRES,ARG1
NSEL,ALL
LSWRITE
*END
*DO,I,0,10
*USE,LOAD,I*38.55
*ENDDO
*DO,I,1,10
*USE,LOAD,385.5-(I*38.55)
*ENDDO
*DO,I,1,10
*USE,LOAD,0.000+(I*38.55)
*ENDDO
*DO,I,1,10
*USE,LOAD,385.5-(I*38.55)
*ENDDO
C***
LSSOLVE,1,41
FINISH
```

Model: 2-dim plastic with crack

```
/TITLE,breech block (2-dim plastic
  w/crack)
/PREP7
C***
ET,1,42,,,2
ET,2,2,,,2
ET,3,12
MPTEMP,1,0,500
MP,EX,1,193E3,1
MP,NUXY,1,.3,0.0001
TB,MKIN,1,2
TBTEMP,,STRAIN
TBDATA,1,0.002590,0.005924,0.079997,0.
  2,0.73
TBTEMP,0.0
TBDATA,1,500,758,1200,1380,1654
TBTEMP,500
TBDATA,1,500,759,1201,1381,1655
R,31,0,0.14E6,1.0E-6
R,32,0,0.065E6,1.0E-6
R,33,-35,1E6,1.0E-6
```

```
C***
C*** NODES
N,001,30.000,117.00,0.0
N,002,30.000,118.27,0.0
N,003,30.000,119.74,0.0
...
...
...
C*** ELEMENTS
E,009,039,040,010
E,010,040,041,011
E,011,041,042,012
...
...
...
ETYPE,2
E,156,157,146
E,325,312,324
...
...
...
TYPE,3
REAL,31
E,457,456
REAL,32
E,404,403
C***
REAL,33
E,1210,210
E,1211,211
C***
WSORT,Y
FINISH
C***
/SOLUTION
C*** BOUNDARY CONDITIONS
D,457,ALL
D,404,ALL
NSEL,S,LOC,X,119.9,120.1
DSYM,SYMM,X,0
NSEL,ALL
C***
*CREATE,LOAD
NSEL,S,LOC,Y,150.
NSEL,R,LOC,X,37.5,120.
SF,ALL,PRES,ARG1
NSEL,ALL
LSWRITE
*END
*DO,I,0,10
*USE,LOAD,I*38.55
*ENDDO
*DO,I,1,10
*USE,LOAD,385.5-(I*38.55)
*ENDDO
*DO,I,1,10
*USE,LOAD,0.000+(I*38.55)
*ENDDO
*DO,I,1,10
*USE,LOAD,385.5-(I*38.55)
*ENDDO
C***
LSSOLVE,1,41
FINISH

***
```

COKE DRUM

```
/TITLE, coke drum
/PREP7
C***
ET,1,13,4,,1
MP,KXX,1,593.248e-6,0.10494e-6
MP,ALPX,,7.68e-6
MP,DENS,1,0.284
MP,C,1,0.1029,0.523e-4
MP,HF,1,0.0,0.0011111! WATER COOL
MP,HF,2,0.085e-6! AIR COOLING
MP,HF,3,243.3e-6,-0.2167e-6! OIL
   HEATIN
MPTEMP,1,80,900
MP,EX,1,30487805,-6097.56
MP,NUXY,1,.3,0.000001
TB,MISO,1,2
TBTEMP,,STRAIN
TBDATA,1,0.00032,0.001,0.003,0.01,0.2
TBTEMP,80.0
TBDATA,1,9500,22000,35000,52000,122000
TBTEMP,900
TBDATA,1,8000,17000,27000,43000,114000
C***
N,1,162.0
N,31,163.26
FILL,,,,,,,,4
NGEN,2,100,1,31,1,,,.025
E,101,1,2,102
EGEN,30,1,1
CP,8,UX,1,101
CPSGEN,31,1,8
C***
FINISH
C***
/SOLUTION
C***
C*** BOUNDARY CONDITIONS
D,1,UY,,,31
NSEL,S,NODE,,1,101,100
SF,ALL,PRES,24.0! PRESSURE IN
NSEL,ALL
NSEL,S,NODE,,101,131,1
SF,ALL,PRES,255.0! WEIGHT
NSEL,ALL
C***
C*** THERMAL LOADING
ANTYPE,TRANS
NEQIT,100
TIMINT,OFF,STRUC
CNVTOL,HEAT
CNVTOL,F
AUTOTS,ON
KBC,1
TREF,80! SET REFERENCE TEM
TUNIF,100! SET INITIAL UNIFO T
NSEL,S,NODE,,1,101,100
SF,ALL,CONV,-3,900.0! OIL HEATING
NSEL,ALL
NSEL,S,NODE,,31,131,100
SF,ALL,CONV,-2,80.0! AIR COOLING
NSEL,ALL
NROPT,1
DELTIM,0.01,,0.60
TIME,20.0! TIME AT END
```

```
LSWRITE                                   NSEL,S,NODE,,1,101,100
C***                                      SF,ALL,CONV,-1,100.0! WATER COOLIN
NSEL,S,NODE,,1,101,100                    NSEL,ALL
SF,ALL,CONV,-3,900.0! OIL HEATING         NSEL,S,NODE,,31,131,100
NSEL,ALL                                  SF,ALL,CONV,-2,80.0! AIR COOLING
NSEL,S,NODE,,31,131,100                   NSEL,ALL
SF,ALL,CONV,-2,80.0! AIR COOLING          DELTIM,0.01,,0.60
NSEL,ALL                                  TIME,3900.! TIME AT END
DELTIM,0.01,,0.60                         LSWRITE
TIME,300.0! TIME AT END                   C***
LSWRITE                                   NSEL,S,NODE,,1,101,100
C***                                      SF,ALL,CONV,-1,100.0! WATER COOLIN
NSEL,S,NODE,,1,101,100                    NSEL,ALL
SF,ALL,CONV,-3,900.0! OIL HEATING         NSEL,S,NODE,,31,131,100
NSEL,ALL                                  SF,ALL,CONV,-2,80.0! AIR COOLING
NSEL,S,NODE,,31,131,100                   NSEL,ALL
SF,ALL,CONV,-2,80.0! AIR COOLING          DELTIM,0.01,,0.60
NSEL,ALL                                  TIME,7200.! TIME AT END
DELTIM,0.01,,0.60                         LSWRITE
TIME,3600.! TIME AT END                   C***
LSWRITE                                   NSEL,S,NODE,,1,101,100
C***                                      SF,ALL,CONV,-3,900.0! OIL HEATING
NSEL,S,NODE,,1,101,100                    NSEL,ALL
SF,ALL,CONV,-1,100.0! WATER COOLIN        NSEL,S,NODE,,31,131,100
NSEL,ALL                                  SF,ALL,CONV,-2,80.0! AIR COOLING
NSEL,S,NODE,,31,131,100                   NSEL,ALL
SF,ALL,CONV,-2,80.0! AIR COOLING          DELTIM,0.01,,0.60
NSEL,ALL                                  TIME,7220.! TIME AT END
DELTIM,0.01,,0.60                         LSWRITE
TIME,3620.! TIME AT END                   C***
LSWRITE                                   LSSOLVE,1,7,1
C***                                      FINISH
```

TABLE A.5
Input Data Listing for Fatigue Cases—MSC/NASTRAN Program

```
FUSE PIN

ID NASTRAN
SOL 106
TIME 999
CEND
$
TITLE=FUSE
MAXLINES=20000000
SET 101=128
DISP=101
SET 102=75
STRESS=102
STRAIN=102
SPC=11
SUBCASE 1
LOAD=1
NLPARM=10
SUBCASE 2
NLPARM=20
SUBCASE 3
LOAD=1
NLPARM=30
$ SUBCASE 4
$ NLPARM=40
BEGIN BULK
PARAM,POST,0
$
$ GRID POINTS
GRID,1,0,0.00000,27.300,0.00000
GRID,2,0,0.00000,19.304,19.304
GRID,3,0,0.00000,26.370,7.0658
...
...
...
$ ELEMENTS
CHEXA,1,1,1,11,13,3,44,49,57,46
CHEXA,2,1,11,12,14,13,49,51,61,57
CHEXA,3,1,12,8,9,14,51,23,25,61
...
...
...
CGAP,865,2,120, 949,1000.,0.,0.
CGAP,866,2,122, 950,1000.,0.,0.
CGAP,867,2,124, 951,1000.,0.,0.
...
...
...
$
$ Boundary conditions
$
FORCE,1,1353,,1409.375,0.,-1.,0.
FORCE,1,1354,,1409.375,0.,-1.,0.
FORCE,1,1358,,1409.375,0.,-1.,0.
...
...
...
SPC,11,1413,13456
SPC,11,1353,13456
SPC,11,1354,13456
```

```
...
...
...
SPC1,11,2,833,834,835,836,837,838
SPC1,11,2,839,840,841,842,843,844
SPC1,11,2,845,846,847,848,849,850
...
...
...
$ Element and material
$ properties
$
MAT1,1,193.E3.,,,.3
MATS1,1,1,PLASTIC,,1,1,193.
TABLES1,1
,0.,0.,0.001,193.0,0.002,320.0,0.005,4
  70.0
,0.015,630.0,0.45,1200.0,ENDT
$
PSOLID,1,1,,,1
PGAP,2,,,1.4E6
$
$ Loading parameters
$
NLPARM,10,3,,,5,25,W,YES
NLPARM,20,3,,,5,25,W,YES
NLPARM,30,3,,,5,25,W,YES
$ NLPARM,40,3,,,5,25,W,YES
$
ENDDATA

***
```

BREECH BLOCK

Model: 3-dim coarse

```
ID NASTRAN
SOL 101
TIME 999
CEND
$
TITLE=BREECH BLOCK 3-dim coarse
DISP=ALL
STRESS=ALL
SPCFORCE=ALL
SPC=200
LOAD=100
BEGIN BULK
PARAM,POST,0
$
$ GRID POINTS
GRID,1,0,0.0,0.0,0.0
GRID,2,0,30.0,0.0,0.0
GRID,3,0,37.5,0.0,0.0
...
...
...
$ ELEMENTS
CHEXA,1,1, 1, 2, 7, 6,101,102,107,106
```

```
CHEXA,2,1, 2, 3, 8, 7,102,103,108,107
CHEXA,3,1, 3, 4, 9, 8,103,104,109,108
...
...
...
CGAP,259,1000,701,601,1000.,1000.,180.
CGAP,260,1000,702,602,1000.,1000.,180.
CGAP,261,1000,703,603,1000.,1000.,180.
...
...
...
$ Boundary conditions
$
SPC,200,11,1
SPC,200,17,1
SPC,200,23,1
...
...
...
SPC1,200,123,701,THRU,757
$
PLOAD4,100,13,731.54,,,,15,22
,,0,0,1
PLOAD4,100,14,731.54,,,,16,23
,,0,0,1
PLOAD4,100,17,731.54,,,,20,27
,,0,0,1
...
...
...
$ Element and Material
$ Properties
$
MAT1,1,193.E3.,,.3
PSOLID,1,1,,,1
PGAP,1000,,,1.4E6
$
ENDDATA
```

Model: 2-dim elastic coarse

```
ID NASTRAN
SOL 101
TIME 999
CEND
$
TITLE = BREECH BLOCK 2-dim elastic
  coarse
ECHO=NONE
SPC=15
LOAD=25
STRESS=ALL
DISP=ALL
BEGIN BULK
PARAM,POST,0
PARAM,K6ROT,1.
$
CORD2R,10,0,0.,0.,0.,0.,0.,0.,1.
,0.,1.,1.,
$
$ GRID POINTS
GRID,1,0,0.0,0.0,0.0
GRID,2,0,30.0,0.0,0.0
GRID,3,0,37.5,0.0,0.0
```

```
...
...
...
$ ELEMENTS
CQUAD4,1,11, 1, 2,12,11
CQUAD4,2,11, 2, 3,13,12
CQUAD4,3,11, 3, 4,14,13
...
...
...
CGAP,27,21,101,1,1000.,1000.,0.
CGAP,28,22,102,2,1000.,1000.,0.
$
$ Boundary conditions
$
SPC1,15,12,101,102
SPC1,15,1,6,16,26,36,46,56
,66
$
GRDSET,,,,,,,3456
$
FORCE,25,63,, 4529.6,,-1.0
FORCE,25,64,, 9155.6,,-1.0
FORCE,25,65,,11372.3,,-1.0
FORCE,25,66,, 6746.3,,-1.0
$
$ Element and Material
$ Properties
$
$
MAT1,1,212.E3,,.42857
PSHELL,11,1,1.0,1
PGAP,21,0.,0.,0.14E6
PGAP,22,0.,0.,0.065E6
$
$
$ Note: this is a plane strain
  problem,
$ therefore we use
$
$ E=193E3/(1 - 0.3**2)=212E3
$ nu=0.3/(1 - 0.3)=0.42857
$
$
ENDDATA
```

Model: 2-dim elastic precise

```
ID NASTRAN
SOL 101
TIME 999
CEND
$
TITLE = BREECH BLOCK 2-dim elastic
  precise
ECHO=NONE
SPC=15
LOAD=25
STRESS=ALL
DISP=ALL
BEGIN BULK
PARAM,POST,0
PARAM,K6ROT,1.
$
```

```
CORD2R,10,0,0.,0.,0.,0.,0.,1.
,0.,1.,1.,
$
$ GRID POINTS
GRID,1,0,30.000,117.00,0.0
GRID,2,0,30.000,118.27,0.0
GRID,3,0,30.000,119.74,0.0
...
...
...
$ ELEMENTS
CQUAD4,1,11,9,39,40,10
CQUAD4,2,11,10,40,41,11
CQUAD4,3,11,11,41,42,12
...
...
...
CTRIA3,388,11,156,157,146
CTRIA3,389,11,325,312,324
CTRIA3,390,11,397,396,414
...
...
...
CGAP,430,21,457,456,1000.,1000.,0.
CGAP,431,22,404,403,1000.,1000.,0.
$
$ Boundary conditions
$
SPC1,15,12,457,404
SPC1,15,1,87,114,131,161,176,198
,228,257,278,283,306,337
$
GRDSET,,,,,,,3456
$
FORCE,25,42,,2280.8,,-1.0
FORCE,25,56,,4561.8,,-1.0
FORCE,25,70,,4561.8,,-1.0
...
...
...
$ Element and Material
$ Properties
$
$
MAT1,1,212.E3,,.42857
PSHELL,11,1,1.0,1
PGAP,21,0.,0.,0.14E6
PGAP,22,0.,0.,0.065E6
$
$
$ Note: this is a plane strain
  problem,
$ therefore we use
$
$ E=193E3/(1 - 0.3**2)=212E3
$ nu=0.3/(1 - 0.3)=0.42857
$
$
ENDDATA
```

Model: 2-dim plastic precise

```
ID NASTRAN
SOL 106
TIME 999
```

```
CEND
$
TITLE = BREECH BLOCK 2-dim plastic
  precise
ECHO=NONE
SPC=15
LOAD=25
STRESS=ALL
STRAIN=ALL
DISP=ALL
NLPARM=27
BEGIN BULK
PARAM,POST,0
$
CORD2R,10,0,0.,0.,0.,0.,0.,1.
,0.,1.,1.,
$
$ GRID POINTS
GRID,1,0,30.000,117.00,0.0
GRID,2,0,30.000,118.27,0.0
GRID,3,0,30.000,119.74,0.0
...
...
...
$ ELEMENTS
CQUAD4,1,11,9,39,40,10
CQUAD4,2,11,10,40,41,11
CQUAD4,3,11,11,41,42,12
...
...
...
CTRIA3,388,11,156,157,146
CTRIA3,389,11,325,312,324
CTRIA3,390,11,397,396,414
...
...
...
CGAP,430,21,457,456,1000.,1000.,0.
CGAP,431,22,404,403,1000.,1000.,0.
$
$ Boundary conditions
$
SPC1,15,12,457,404
SPC1,15,1,87,114,131,161,176,198
,228,257,278,283,306,337
$
GRDSET,,,,,,,3456
$
FORCE,25,42,,2280.8,,-1.0
FORCE,25,56,,4561.8,,-1.0
FORCE,25,70,,4561.8,,-1.0
...
...
...
$ Element and Material
$ Properties
$
MAT1,1,212.E3,,.42857
MATS1,1,1,PLASTIC,,1,1,450.
TABLES1,1
,0.,0.,0.002590,500.,0.005924,758.,0.0
  79997,1200.
,0.2,1380.,0.73,1654.,ENDT
$
PSHELL,11,1,1.0,1
```

```
PGAP,21,0.,0.,0.14E6                    ...
PGAP,22,0.,0.,0.065E6                   ...
$                                       ...
$ Loading parameter                     CQUAD4,232,1,200, 183, 184,1002
$                                       CQUAD4,233,1,184, 185, 202,1003
NLPARM,27,10,,,,,W,YES                  CQUAD4,240,1,211, 210, 199,1001
$                                       ...
ENDDATA                                 ...
                                        ...
```

Model: 2-dim plastic with crack

```
                                        CTRIA3,1001,1,199,200,1001
ID NASTRAN                              CTRIA3,1002,1,184,201,1002
SOL 106                                 CTRIA3,1003,1,201,184,1003
TIME 999                                ...
CEND                                    ...
$                                       ...
TITLE = BREECH BLOCK 2-dim plastic      CRAC2D,2000,1000,212,211,200,201,202,2
  with crack                             13
ECHO=NONE                               ,242,241,240,1211,1001,1002,1003,1004
SPC=15                                  ,1005,1006,1007,1008
LOAD=25                                 ADUM8,18,0,5,0,CRAC2D
STRESS=ALL                              $
STRAIN=ALL                              $ Boundary conditions
DISP=ALL                                $
NLPARM=27                               SPC1,15,12,457,404
BEGIN BULK                              SPC1,15,1,87,114,131,161,176,198
PARAM,POST,0                            ,228,257,278,283,306,337
$                                       $
CORD2R,10,0,0.,0.,0.,0.,0.,1.           GRDSET,,,,,,3456
,0.,1.,1.,                              $
$                                       FORCE,25,42,,2280.8,,-1.0
$                                       FORCE,25,56,,4561.8,,-1.0
$ The model is obtained from the        FORCE,25,70,,4561.8,,-1.0
  plastic                               FORCE,25,84,,5300.6,,-1.0
$ precise model after the following     ...
$ revisions:                            ...
$                                       $ Element and Material
$ 1. CQUAD4 elements 241, 242, 252 and  $ Properties
$ 253 are deleted.                      $
$ 2. Grids 1001 to 1008 are added, see  MAT1,1,212.E3,,.42857
$ below.                                MATS1,1,1,PLASTIC,,1,1,450.
$ 3. CQUAD4l elements 232, 233, 240,    TABLES1,1
  243,                                  ,0.,0.,0.002590,500.,0.005924,758.,0.0
$ 251, 254, 265 and 264 are revised,      7999,1200.
  see                                   ,0.2,1380.,0.73,1654.,ENDT
$ below.                                $
$ 4. CTRIA3 elements 1001 to 1008 are   PSHELL,1,1,1.0,1
  added,                                PGAP,21,0.,0.,0.14E6
$ see below.                            PGAP,22,0.,0.,0.065E6
$ 5. Crack element 2001 is added, see   PRAC2D,1000,1,1.,0
  below.                                $
$                                       $ Loading parameter
$                                       $
GRID,1001,0,37.883,95.907,0.            NLPARM,27,10,,,,,W,YES
GRID,1002,0,38.385,95.978,0.            $
GRID,1003,0,39.168,95.563,0.            ENDDATA
```

NOMENCLATURE

a	crack length; empirical factor; exponent
a_{eff}	effective crack length
a_f	final crack length
a_i	initial crack length; relative number of cycles
b	fatigue strength exponent; material constant; width
c	fatigue ductility exponent; specific heat
d	diameter; size of stress field
e	engineering strain
e_{ij}	strain tensor
$e^{(1)}_{ij}, e^{(2)}_{ij}, e^{(3)}_{ij}$	component strain tensors
f	feed
$f(\ldots)$	yield criterion
$f_{\text{I}}, f_{\text{II}}, f_{\text{III}}$	compliance functions
g	gravitational constant
h	heat transfer coefficient
h_s	strain hardening depth
k	mean stress sensitivity coefficient; thermal conductivity
k_a, k_b, k_c	fatigue limit modification factors

k_d	fatigue limit reliability factor
l	length
l_o	original length of specimen
n	normal direction; strain hardening exponent
n_i	number of load cycles at load level i
$n_1, n_2, \ldots$	number of load cycles at load levels 1, 2, ...
n'	cyclic strain hardening exponent
p	pressure
q	notch sensitivity
r	radius
$r_1, r_2, \ldots$	locations of strain energy density factors
s_{ij}	stress deviator tensor
s_x, s_y, s_z	stress components
s_1, s_2, s_3	principal stress deviators
t	time
u, v, w	displacement components in x, y, z directions
u_i	displacements
$\{\mathbf{u}\}$	displacement vector
$\{\mathbf{u_e}\}$	vector of nodal displacements in an element
$\{\mathbf{u_g}\}$	global vector of nodal displacements
x, y, z	directions in Cartesian coordinate system
A	area; ratio of stress amplitude to mean stress
A_f	cross-section area at fracture
A_o	original cross-section area
A_r	average spacing of roughness peaks
B	thickness of impact test specimen; wall thickness
BHN	Brinell hardness number
$[\mathbf{B}]$	strain matrix
C	compliance function; constant factor
D	diameter; notch size
D_i	relative damage
$[\mathbf{D}]$	elasticity matrix
$[\mathbf{D_e}]$	elasticity matrix
$[\mathbf{D_{ep}}]$	plasticity matrix
E	modulus of elasticity
F	reliability
$F(\ldots)$	yield function
FS	factor of safety
FS'	supplementary factor of safety
$\{\mathbf{F_e}\}$	vector of nodal forces in element
G	shear modulus; elastic energy release rate

G_c	critical value of elastic energy release rate
G_s	strain hardening gradient
G_r	Grashoff number of heat transfer
G_I, G_{II}, G_{III}	elastic energy release rates
HV	Vickers hardness
H_μ	microhardness
I	moment of inertia
I_1, I_2, I_3	stress invariants
J	elastic-plastic energy release rate; polar moment of inertia
J_1, J_2, J_3	stress deviator invariants
K	strength coefficient; stress intensity factor
K_c	fracture toughness
K_f	fatigue notch factor
K_{th}	threshold stress intensity factor
K_t	stress concentration factor
K_σ	true stress concentration factor
K_ε	true strain concentration factor
$K_{I,eq}$	equivalent stress intensity factor
K_{Ic}	fracture toughness
K_I, K_{II}, K_{III}	stress intensity factors
K'	cyclic strength coefficient
K'_f	fatigue notch factor for limited life
$[K_e]$	element stiffness matrix
$[K_g]$	global stiffness matrix
$[K_{ep}]$	elastic-plastic stiffness matrix
$[K_{gn}]$	global nonlinear stiffness matrix
M	bending moment
N	life expectancy; number of cycles to failure
N_i	life expectancy at load level i; shape function components
N_f	number of cycles to failure
$N_1, N_2, \ldots$	life expectancy at load levels 1, 2, ... ; shape function components
$[N]$	shape function
P	cumulative frequency; force; probability
R	radius; ratio of minimum stress to maximum stress
R_a, R_p, R_q, R_y, R_z	surface roughness parameters
R_c	Rockwell C hardness
R_{max}	maximum peak-to-valey roughness height
Ra	Rayleigh number of heat transfer
$\{R_g\}$	vector of external forces
S	strain energy density factor; stress; engineering stress

S_a	stress amplitude
$S_{a,\text{lim}}$	limit stress amplitude
S_{ax}, S_{ay}, S_{az}	stress amplitude components
S_{a1}, S_{a2}, S_{a3}	principal stress amplitudes
S_f	fatigue limit
S_c	critical value of strain energy density factor
S_m	mean stress
S_{mx}, S_{my}, S_{mz}	mean stress components
S_{m1}, S_{m2}, S_{m3}	principal mean stresses
S_{max}	maximum stress
S_{min}	minimum stress
S_N	fatigue strength at life expectancy N
S_r	stress range
$S_{r,\text{lim}}$	limit stress range
$S_{x,a}, S_{y,a}, S_{z,a}$	same as S_{ax}, S_{ay}, S_{az}
$S_{x,m}, S_{y,m}, S_{z,m}$	same as S_{mx}, S_{my}, S_{mz}
S_u	ultimate tensile strength
S_y	yield point
$S_1, S_2, \ldots$	values of strain energy density factor
S'_a	equivalent stress amplitude
S'_f	fatigue limit of standard specimen
S'_m	equivalent mean stress
S''_f	fatigue limit of specimen at a specified reliability
T	temperature; torque
T_i	external forces
U	elastic strain energy
U_o	elastic strain energy per unit volume
U_s	strain-hardening rate
V	volume
W	width of impact test specimen; work
W_F	work performed by external forces
W_i	deformation work
W_{io}	deformation work per unit volume
α	crack propagation factor; thermal expansion coefficient
γ_e	equivalent shear strain
$\gamma_{xy}, \gamma_{yz}, \gamma_{zx}$	shear strain components
$\gamma^p_{xy}, \gamma^p_{yz}, \gamma^p_{zx}$	plastic shear strain components
ε	strain; true strain
ε_e	elastic strain; equivalent elastic strain
ε_f	true fracture ductility
ε_{ij}	strain tensor; strains

ε_p	plastic strain; equivalent plastic strain
$\varepsilon_x, \varepsilon_y, \varepsilon_z$	normal strain components
ε_{yp}	strain at yield point
$\varepsilon_1, \varepsilon_2, \varepsilon_3$	principal strain components
ε^e_{ij}	elastic strains
$\varepsilon^e_1, \varepsilon^e_2, \varepsilon^e_3$	elastic principal strain components
ε^p_{ij}	plastic strains
$\varepsilon^p_1, \varepsilon^p_2, \varepsilon^p_3$	plastic principal strain components
ε'_f	fatigue ductility coefficient
$\{\varepsilon\}$	strain vector
θ	polar angle in crack coordinate system
κ	parameter defined by Equations (5.9) and (5.10); plastic deformation parameter
λ	flow-rule proportionality factor
ν	Poisson ratio
ρ	density
σ	standard deviation; stress; true stress
σ_e	critical stress
σ_e	equivalent stress
σ_f	true fracture strength
σ_{ij}	stress tensor; stresses
σ_m	mean stress
$\sigma_{m,e}$	equivalent mean stress
σ_{nom}	nominal stress
σ_N	equivalent stress amplitude
σ_r	residual stress
$\sigma_x, \sigma_y, \sigma_z$	normal stress components;
σ_{yp}	yield point
$\sigma_1, \sigma_2, \sigma_3$	principal stress components
$\sigma_{1,m}, \sigma_{2,m}, \sigma_{3,m}$	principal mean stress components
σ'_f	fatigue strength coefficient
τ_e	equivalent shear stress
τ_f	fatigue limit in shear
$\tau_{axy}, \tau_{ayz}, \tau_{azx}$	alternating shear stress components
$\tau_{xy}, \tau_{yz}, \tau_{zx}$	shear stress components
$\tau_{xy,a}, \tau_{yz,a}, \tau_{zx,a}$	same as $\tau_{axy}, \tau_{ayz}, \tau_{azx}$
$\tau_{12}, \tau_{23}, \tau_{31}$	maximum shear stresses

INDEX

ABOUT THE
AUTHORS

ELIAHU ZAHAVI is professor emeritus at the Ben-Gurion University in Israel, where for the past twenty-five years he has been active both as a teacher and a researcher. He joined Ben-Gurion University after obtaining a D.Sc. He was responsible for developing the machine design group at Ben-Gurion. Prior to that he was involved in aero-space industry as a design engineer. His special interest is in the application of computers in machine design. He is the author of *The Finite Element Method In Machine Design* and numerous papers on the topics of engineering science and engineering education.

VLADIMIR TORBILO is a former professor of the Perm Technical University in Russia. He now resides in Israel where he is teaching mathematics and is devoting his time to writing. He has a D.Sc. degree in mechanical engineering from Moscow Automotive Institute and was director of the Laboratory of Finishing Methods in Machine Manufacturing. His expertise is the improvement of machine parts reliability. He is the author of *Diamond Burnishing* (Moscow, 1972) and has published more than a hundred scientific papers.

Milton Keynes UK
Ingram Content Group UK Ltd.
UKHW052019071024
449327UK00027B/2344